Lecture Notes in Electrical Engineering

Volume 635

The book series *Lecture Notes in Electrical Engineering* (LNEE) publishes the latest developments in Electrical Engineering - quickly, informally and in high quality. While original research reported in proceedings and monographs has traditionally formed the core of LNEE, we also encourage authors to submit books devoted to supporting student education and professional training in the various fields and applications areas of electrical engineering. The series cover classical and emerging topics concerning:

- Communication Engineering, Information Theory and Networks
- Electronics Engineering and Microelectronics
- Signal, Image and Speech Processing
- Wireless and Mobile Communication
- Circuits and Systems
- Energy Systems, Power Electronics and Electrical Machines
- Electro-optical Engineering
- Instrumentation Engineering
- Avionics Engineering
- Control Systems
- Internet-of-Things and Cybersecurity
- Biomedical Devices, MEMS and NEMS

For general information about this book series, comments or suggestions, please contact leontina.dicecco@springer.com.

To submit a proposal or request further information, please contact the Publishing Editor in your country:

China

Jasmine Dou, Associate Editor (jasmine.dou@springer.com)

India, Japan, Rest of Asia

Swati Meherishi, Executive Editor (Swati.Meherishi@springer.com)

Southeast Asia, Australia, New Zealand

Ramesh Nath Premnath, Editor (ramesh.premnath@springernature.com)

USA, Canada:

Michael Luby, Senior Editor (michael.luby@springer.com)

All other Countries:

Leontina Di Cecco, Senior Editor (leontina.dicecco@springer.com)

**** Indexing: The books of this series are submitted to ISI Proceedings, EI-Compendex, SCOPUS, MetaPress, Web of Science and Springerlink ****

More information about this series at http://www.springer.com/series/7818

Haifeng Li · Shengchen Li · Lin Ma ·
Chunying Fang · Yidan Zhu
Editors

Proceedings of the 7th Conference on Sound and Music Technology (CSMT)

Revised Selected Papers

Springer

Editors
Haifeng Li
School of Computer Science
and Technology
Harbin Institute of Technology (HIT)
Harbin, Heilongjiang, China

Lin Ma
School of Computer Science
and Technology
Harbin Institute of Technology (HIT)
Harbin, Heilongjiang, China

Yidan Zhu
The acoustical society of Beijing
Xicheng, Beijing, China

Shengchen Li
Beijing University of Posts
and Telecommunications
Beijing, China

Chunying Fang
School of Computer and Information
Engineering
Heilongjiang University of science
and technology
Harbin, Heilongjiang, China

ISSN 1876-1100 ISSN 1876-1119 (electronic)
Lecture Notes in Electrical Engineering
ISBN 978-981-15-2755-5 ISBN 978-981-15-2756-2 (eBook)
https://doi.org/10.1007/978-981-15-2756-2

This Springer imprint is published by the registered company Springer Nature Singapore Pte Ltd.
The registered company address is: 152 Beach Road, #21-01/04 Gateway East, Singapore 189721, Singapore

Preface

It is with the great pleasure that Conference on Sound and Music Technology (CSMT) has published its second proceeding with Springer. The conference this year is held in Harbin by Harbin Institution of Technology (HIT). This proceeding contains ten selected papers from submissions and one invited paper ranging from acoustic signal processing to computational musicology.

With the first attempt to publish its own proceeding last year, the 6th edition of CSMT has gained a great success in terms of submissions. This year the 7th edition of CSMT has even collected more submissions than the last year. These numbers clearly document the progress that CSMT has made and give a clear sign of the growth of music technology society in China.

In the past year, music technology has rapid developed, several research projects that combines artificial intelligence and music have emerged. The terms techniques and music are no longer two separate terms that represents science and arts. The encounter of science and art not only provides new source of creativity but also provides a deeper but more objective understanding of music concepts.

The articles in this proceeding cover the research on different types of signals include speech, acoustic and music signal. Besides traditional information retrieval where basic semantic concepts are retrieved from signals, the understanding and perception of signals has also been investigated by the researchers.

The contributors of the proceeding come from multiple universities in China and for the first time from several famous oversea universities. This fact reflects that Chinese researchers in sound and music technology are currently working globally and also remarks that CSMT starts to expand beyond a domestic conference.

Delighted and excited, we, the editors of CSMT proceeding, anticipate the rapid growth of CSMT conference and the Chinese sound and music technology community in Chinese. We welcome anybody to submit in future CSMT events and hope to meet researchers from different countries in China.

Contents

Computational Musicology

Acoustic Signal Processing

Part I
Speech Processing

Bandwidth Extension WaveNet for Bone-Conducted Speech Enhancement

Changyan Zheng, Jibin Yang, Xiongwei Zhang, Tieyong Cao, Meng Sun, Linlin Zheng
Army Engineering University of PLA, 210007

Abstract. Bone-conducted (BC) speech is immune to background noise, but sounds muffled due to the characteristic of human body channel. The existing enhancement methods about BC speech mainly focus on the enhancement of the magnitude spectrum and ignore processing the mismatched phase. Besides, most of the existing communication systems of BC speech are based on low sampling rate configuration and cannot meet the better speech quality demand of the wideband speech communication. In this paper, a novel waveform generation method based on Bandwidth Extension WaveNet (BE-WaveNet) model is proposed, that builds the probability density between the conditional acoustic feature and the target waveform using the deep convolutional neural networks WaveNet and is able to generate the waveform from the enhanced magnitude spectrum directly. In order to further improve the speech quality, an upsampling module of cross resampling rate is introduced, enabling the WaveNet to generate high-sampling rate speech with low-sampling rate acoustic feature, namely the BE-WaveNet is able to extend the bandwidth without increasing the communication cost. The experimental results show that compared with the waveform synthesis methods such as using the original phase and the phase estimation method of Griffin-Lim, the proposed method significantly improves the quality of BC speech. At the same time, high-sampling rate enhanced speech can be obtained without complex system configuration updates.

Keywords: Bone-conducted speech enhancement, Waveform generation, WaveNet, Bandwidth extension.

1 Introduction

Bone-conducted (BC) microphone is a kind of skin-attached non-audible sensor and converts the vibration of the human body such as throat and skull into electrical signal [1]. BC speech is immune to ambient noise, but does not sound natural and clear like conventional air-conducted (AC) speech. Due to the attenuation of human body channel, BC speech faces severe loss of high-frequency components that are usually higher than 2kHz [2]. Besides, some phonemes like unvoiced fricatives are totally lost, which are generated in the oral or nasal cavity rather than the vocal cord.

BC speech enhancement can be modelled as a voice conversion problem that the BC speech is converted to its corresponding clean AC speech. Early algorithms usually decompose speech signals into spectral envelope features and excitation features

H. Li et al. (eds.), *Proceedings of the 7th Conference on Sound and Music Technology (CSMT)*, Lecture Notes in Electrical Engineering 635,
https://doi.org/10.1007/978-981-15-2756-2_1

based on the source-filter model. Compared with the excitation feature, humans are more sensitive to the spectral envelope feature [2], so most of the algorithms focus on the conversion of low-dimensional spectral envelope. Gaussian mixture models (GMM) and shallow neural network are often used to learn the mapping relationship of envelopes [4]-[7]. Then the enhanced waveform can be synthesized using the original excitation and the converted spectral envelope. With the development of deep neural networks (DNN), it is possible to learn the complex mapping function of high-dimensional spectra directly. In our previous work [8], we explored an effective conversion model based on Bidirectional Long Short-Term Memory Recurrent Neural Networks (BLSTM-RNN), which maps the log Short-Time Fourier Transformation (STFT) magnitude spectrum from BC speech to that of AC speech and a great improvement is achieved.

After obtaining the converted spectral envelope or magnitude spectrum, the above methods usually use the original excitation or phase of BC speech to synthesize the waveform. In fact, the excitation and phase are mismatched with the converted features, which causes distortions when synthesizing the speech. In the field of speech de-noising and text-to-speech (TTS), researchers usually use phase estimation algorithms such as Griffin-Lim [9] to relieve the phase mismatch problem. In recent years, a few researchers try to use DNN to estimate phase [10]. The state-of-the art waveform synthesis methods are based on the direct waveform modelling model, such as WaveNet [11] and SampleRNN [12]. They can model the probability density of speech waveform samples by constructing special neural networks and is able to generate speech waveform from the given acoustic features directly. Compared with phase estimation methods, the problems of speech overlapping and parametric over-smoothing are avoided, which make a great breakthrough in speech signal processing tasks. At the same time, most of BC speech communication systems are based on 8kHz sampling rate configuration. In order to acquire high quality speech with higher sampling rate and save the communication costs, we consider to extend the bandwidth of speech in the synthesizer. In fact, some advanced works about bandwidth extension of speech are also based on DNN, such as using generative adversarial networks (GAN) [13] and WaveNet[14].

Inspired by the mentioned works above, we propose a new waveform generation method based on WaveNet for BC speech enhancement, aiming to solve the speech quality degradation caused by the phase mismatch. Meanwhile, WaveNet is also considered to extend the bandwidth of speech to save the cost of communication system updates. To achieve this goal, we introduce an up-sampling module of cross sampling rate in WaveNet, so the enhanced speech with high-sampling rate can be generated by just using the enhanced magnitude spectrum with low-sampling rate. We call the modified WaveNet as Bandwidth Extension WaveNet (BE-WaveNet).

The rest of this paper is organized as follows. The details of the proposed method is introduced in Section 2. A set of evaluation experiments to assess the performance of proposed method are provided in Section 3. Finally, the conclusion is made in Section 4.

2 The Proposed Method

2.1 The Overall Framework

We follow the WaveNet work of TTS [15] and voice conversion [16] and propose the BC speech enhancement framework as shown in Fig. 1. The BLSTM-RNN is used to enhance the magnitude spectrum of BC speech based on our previous work [8], and WaveNet is designed to act as a vocoder. Due to the huge computational complexity, it is not easy for WaveNet to model long-term contextual information, which is quite necessary for inferring the lost components in BC speech. Therefore, the proposed framework has the advantage that it can utilize the powerful contextual modelling ability of BLSTM-RNN to obtain accurate magnitude spectrum, then acquire high quality speech waveform based on the enhance magnitude spectrum via WaveNet. The proposed framework includes a training stage and an enhancement stage.

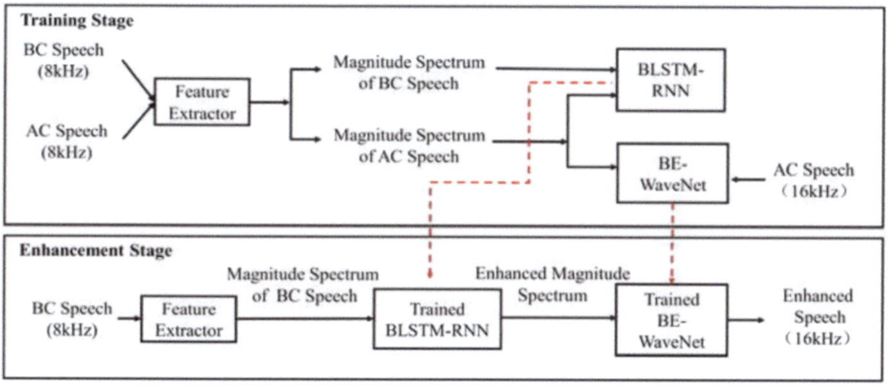

Fig. 1. The Overall framework of BC speech enhancement system.

In the training stage, we firstly extract STFT magnitude, then employ BLSTM-RNN as the magnitude spectrum enhancement model, where the magnitude spectra of BC speech and AC speech act as the input and target respectively. The magnitude spectrum and waveform of AC speech are sent as the input of BE-WaveNet. Since BE-WaveNet is an autoregressive model, the target of it is also the waveform of AC speech.

In the enhancement stage, we firstly extract the STFT magnitude of BC speech to be enhanced, then send the magnitude spectrum to the trained BLSTM-RNN to acquire enhanced magnitude spectrum. Finally, the enhanced spectrum is fed to the trained BE-WaveNet to synthesize the waveform. It can be noticed that the proposed method can obtain enhanced speech with high sampling rate from the feature of speech with low sampling rate, showing the bandwidth extension ability of the proposed BE-WaveNet.

The BLSTM-RNN model originates from our previous work [8], while the BE-WaveNet model is the focus of this paper and will be introduced in the next subsection.

2.2 Bandwidth Extension WaveNet

WaveNet is an autoregressive generative model which can directly model speech waveform. In order to generate waveform with specific properties, it often needs additional conditional input. Assuming the joint probability of a waveform is factorized as a product of conditional probabilities under condition λ:

$$p(x \mid \lambda) = \prod_{t=1}^{T} p(x_t \mid x_1, \ldots, x_{t-1}; \lambda) \tag{1}$$

WaveNet implements the calculation of (1) by stacking well-designed convolutional layers similar with PixelCNN [17]. The architecture of our proposed BE-WaveNet is shown in Fig. 2 and the key components in it will be introduced respectively.

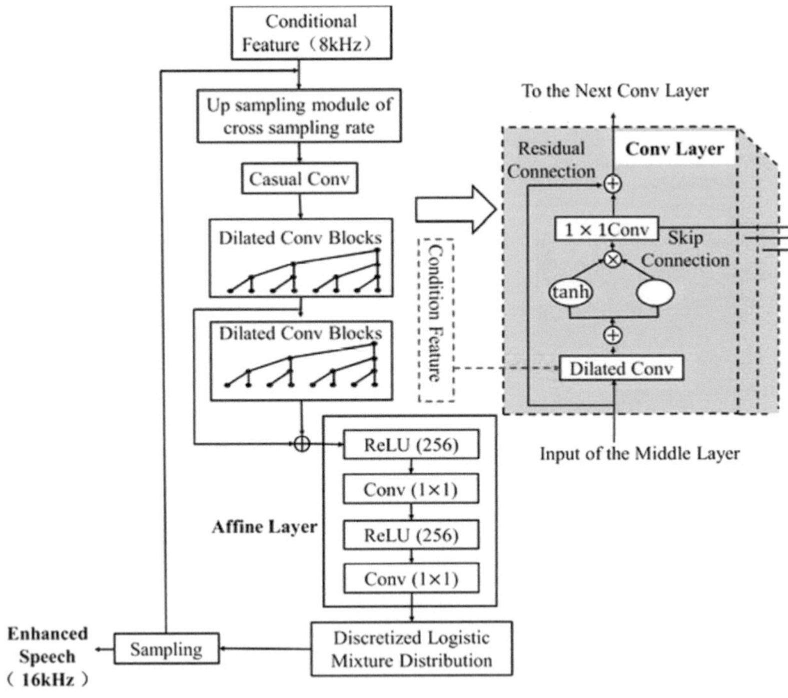

Fig. 2. The Bandwidth Extension WaveNet.

2.2.1 Up-sampling Module of Cross Sampling Rate. The resolution of local conditional features is far lower than the speech waveform, because the former is frame-based feature, while the latter is sample-based. Therefore, the up-sampling module is necessary to adjust the resolutions of the two kinds of features to be the same in WaveNet. Furthermore, in the proposed model, the mismatch of sampling rate of the input and target should also be considered. Therefore, the up-sampling module of cross sampling rate is proposed, which can adjust the resolutions of the local condition features and waveform with different sampling rate.

The module is shown in Fig. 3. The frame-based feature with different sampling rate is extracted with same window time and same window shift, so the frame updating rate s_{low} for all the frames are the same. The linear interpolation method is adopted and the interpolation feature $\Delta e_i = (e_{i+1} - e_i) / M$ is computed according to the difference between the adjacent frame features, where M is the number of samples in a frame. s_i^n is the corresponding speech sampler that i denotes the index of frame and n denotes the index of sampler in the frame.

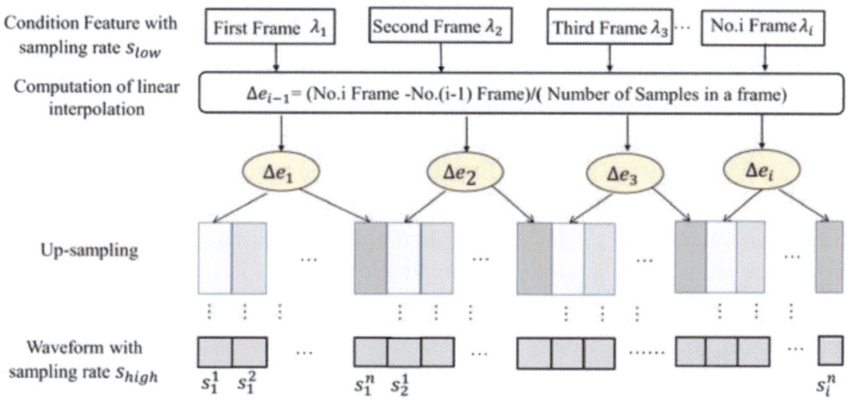

Fig. 3. Up-sampling Module of Cross Sampling Rate.

2.2.2 Blocks of Dilated Convolution. Dilated convolution [18] is used to enlarge the respective field in WaveNet, which can enlarge the size of respective filed exponentially by just increasing the expansion factor of filter size linearly. WaveNet doubles the expansion factor layer by layer to form a block and then stacks these blocks to further increase the receptive field.

Gated Activation Unit (GAU) is employed as the activation function as shown in Fig. 2:

$$z = \tanh(W_{f,k} * x + V_{f,k} * \lambda) \text{ e } \sigma(W_{g,k} * x + V_{g,k} * \lambda) \tag{2}$$

where $*$ and e denote convolution and point-wise multiplication respectively, $\sigma(\cdot)$ is sigmoid function, k is the index of layer, f and g denote filter and gate respectively, W and V denote learnable convolutional filters, z is the output of GAU. This non-linear activation is superior to some other non-linear functions and is a key component to enable WaveNet model the waveform directly.

2.2.2 Discretized Logistic Mixture Distribution. The original output layer of WaveNet is Softmax classifier and the speech signal often needs to be quantified from 16 bits to 8 bits, because it is not possible for Softmax to classify $2^{16} = 65536$ categories. Moreover, Softmax cannot reflect the intrinsic correlation between the different categories, for instance, a value of 128 is close to a value of 127 and 129, but far way with 1. In our method, we deploy Discretized Logistic Mixture Distribution (DLMD) originated from PixelCNN++ [19] instead:

$$v \sim \sum_{i=1}^{K} \alpha_i \text{logistic}(\mu_i, s_i) \tag{3}$$

where K represents the number of Logistic Distribution, α_i and (μ_i, s_i) denote the weight and the parameters of the i-th logistic distribution.

This distribution can simulate the continuous distribution and is more appealing to the real speech distribution. Therefore, the prediction accuracy is improved greatly. In addition, the neural networks predict the parameters α_i, μ_i and s_i, 10 logistic distributions are enough to model the distribution of speech, so totally $3 \times 10 = 30$ values need to be predicted. Therefore, the computation memory is also saved.

3 Experiment

3.1 Dataset and Evaluation Metrics

We collect speech data spoken by 3 female and 3 male speakers using Adobe Audition Software[1] in an anechoic chamber, where the clean speech signal can be acquired. The speech is recorded with 32kHz sampling rate and 16-bit quantization. In order to ensure the synchronization of BC and AC speech, a stereo sound card is used and the two kinds of microphones are connected to the left and the right channel respectively. 1000 Chinese Mandarin sentences are selected as the corpus, and each of them lasts for about 3-5 seconds. All the speakers are required to read 200 different sentences selected randomly in the corpus, thus a parallel BC speech and AC speech dataset of each speaker can be acquired. For validating the methods, 160 sentences are used for training and the rest are for testing, and no overlap exists between the training set and testing set.

[1] Adobe Audition Software is a digital audio workstation from Adobe Systems and is widely used in audio processing.

Mean Opinion Score (MOS) 20 is employed to measure the performance of the proposed method subjectively. 10 professional listeners participated in our testing and they are requested to score the quality of the given speech from 0 to 5 with interval 0.5. The objective evaluation metrics, Short-Time Objective Intelligibility (STOI) 21 and Log-spectral Distance (LSD) 22, are also employed. STOI score ranges from [0, 1], higher STOI scores means better speech intelligibility. LSD is used to measure the spectral distortion between the referenced and enhanced speech. The smaller LSD scores, the better speech quality.

3.2 Experimental Setup and Baseline System

3.2.1 Experimental Setup.

(1) Settings of Feature Extraction

In our proposed scheme, the speech at 8kHz and 16kHz sampling rate is required. Since the collected signal is sampled at 32kHz, all the speech data is firstly down-sampled to 8kHz and 16kHz respectively, and is then windowed by 32ms with 8ms frame shift. 256-point and 512-point STFT are used to extract the magnitude spectrum of speech with 8kHz and 16kHz sampling rate respectively.

(2) Settings of BLSTM-RNN

The magnitude spectrum enhancement module includes 3 layers of BLSTM-RNN, each with 256 hidden units. Batch Normalization (BN) is subsequently used after each hidden layer. The input and target are features of BC speech and AC speech with 8kHz sampling rates respectively.

(3) Settings of BE-WaveNet

BE-WaveNet includes 2 blocks of dilated convolution, each of the blocks contains 8 layers. Therefore, totally 19 layers in the BE-WaveNet with additional 1 casual convolution layer and 2 feature affine layers. The initial kernel size of casual convolution and dilated convolution is set to $3*1$, and the number of kernel is set to 256. The input and target of BE-WaveNet are enhanced BC magnitude spectrum with 8kHz sampling rate and AC speech with 16kHz sampling rate.

3.2.2 Baseline System. To show the effectiveness of the proposed method (denoted as **BE-WaveNet**), we compare it with 4 different waveform synthesis methods, including using the original phase of BC speech method (denoted as **iSTFT**), two phase estimation methods Griffin-Lim and one of its modified version (denoted as **Griffin-Lim** and **Lws**), and conventional WaveNet method (denoted as **WaveNet**). All of the methods are set to acquire enhanced speech at 16kHz sampling rate. It can be concluded that the effectiveness of different vocoders by comparing the 4 methods, since they are given the same sampling rate condition. We have to point out that the magnitude spectrum enhancement model in the 4 methods is still based on BLSTM-RNN, but the number of hidden units each layer is enlarged from 256 to 512 which is different from the proposed method. At the same time, we can compare the proposed **BE-WaveNet** with **WaveNet** to verify its ability of bandwidth extension, because the former is given the enhanced magnitude spectrum at 8kHz. More details of the four methods are described as followings.

(1) **iSTFT**

Synthesizing the waveform by inverse STFT, where original BC phase and enhanced magnitude spectrum are used. This is also the method used in our previous works 8.

(2) **Griffin-Lim**

It is a phase estimation method based on the continuity of STFT coefficients in real values. Its key idea is to estimate the signal with real values and make the values approaching to the given magnitude by minimizing the least square error iteratively.

(3) **Lws**

It is a modified version of Griffin-Lim method, which uses Local Weighted Sums 23 to initialize the phase estimation. Its effectiveness is based on the continuity of STFT coefficients in the complex field.

(4) **WaveNet**

We replace the up-sampling module of cross sampling rate with normal up-sampling module in BE-WaveNet, therefore the model is changed to the conventional WaveNet model. The input of WaveNet is enhanced spectrum magnitude at 16kHz sampling rate.

3.3 Results and Analysis.

(1) Subjective Results

The MOS values of each method are shown in Fig. 4. The 95% confidence intervals are also shown with numerics and I-bars. We can note that the MOS value of original BC speech is very low. WaveNet-based methods show obvious advantages over the other methods, demonstrating the effectiveness of direct waveform modelling methods. The MOS value of BE-WaveNet is close to the value of WaveNet, which indicates that BE-WaveNet can obtain similar speech quality with WaveNet, proving its good bandwidth extension capability. Listeners evaluate the speech enhanced by Griffin-Lim and Lws methods. They believe that the speech contains more mechanical noise and thus their MOS value is low. The MOS value of Lws is much lower than the other methods, we infer it is because Lws estimates the phase according to the continuity of spectro-temporal points, therefore some unexpected structures may also be generated. Since unexpected harmonic structures can make speech sound muffled, the MOS value of Lws is much low.

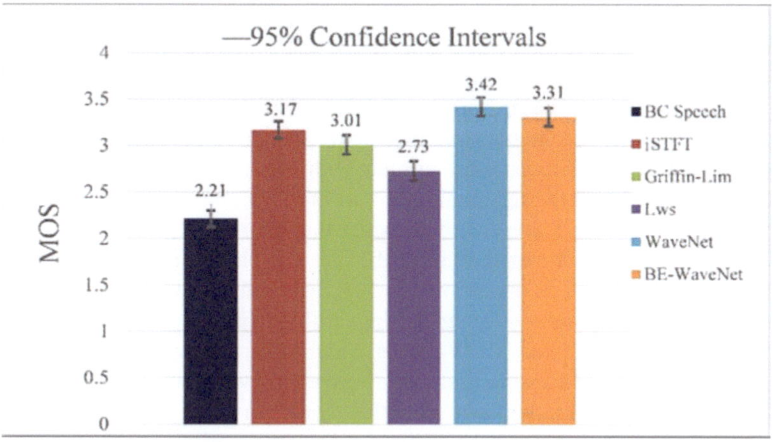

Fig. 4. MOS values of speech enhanced by different methods.

(2) Objective Results

The objective results are presented in Tab. 1.

Table 1. Objective results of different methods.

Person	BC		iSTFT		Griffin-Lim		Lws		WaveNet		BE-WaveNet	
	STOI	LSD	STOI	LSD	STOI	LSD	STOI	LSD	STOI	LSD	STOI	LSD
female1	0.666	1.489	0.897	1.020	0.792	2.062	0.861	0.991	0.891	1.051	0.884	1.084
female2	0.588	1.459	0.830	1.141	0.730	2.121	0.811	1.042	0.846	1.092	0.835	1.104
female3	0.566	1.405	0.800	1.068	0.691	2.150	0.777	1.047	0.812	1.122	0.796	1.135
male1	0.632	1.733	0.849	1.253	0.655	2.433	0.832	1.055	0.795	1.247	0.781	1.261
male2	0.598	1.370	0.839	1.081	0.680	1.985	0.816	1.011	0.832	1.102	0.818	1.119
male3	0.622	1.629	0.796	1.128	0.654	2.458	0.780	1.060	0.812	1.152	0.796	1.164
Average	0.612	1.514	**0.835**	1.115	0.700	2.202	0.813	**1.034**	0.831	1.128	0.884	1.084

From Tab. 1, it can be seen that iSTFT method gets the best STOI score, Lws performs best in LSD metric, while WaveNet-based methods do not show better performance. It is because the optimizing target of conventional waveform synthesis method is mainly based on spectral distance, which is very consistent with existing objective metrics of speech quality, while the optimizing target of WaveNet-based method is the distance of waveform. Therefore, the WaveNet-based methods are not easy to perform better than the traditional methods in objective metrics, but can show great advantages in the subjective metrics.

(3) Analysis of Magnitude Spectrogram

We then analyze the magnitude spectrogram of each method as shown in Fig. 5. In Fig. 5(a), we can see that the frequency components of BC speech is no more than 2.5kHz, while its corresponding AC speech has rich components until 8kHz as shown in Fig. 5(b). From the rectangular and oval boxes, it can be seen that WaveNet-based method can obtain magnitudes more similar with the magnitude of AC speech than the other methods. From the rectangular box in Fig. 5(c), it can be noted that the harmonic structures of iSTFT are distorted, while the left methods have relatively clear structures, which demonstrates that using the original phase of BC speech to synthesize the speech can result in distortions. In Fig. 5(d) and (e), we can see that Griffin-Lim and Lws meet obvious over-smoothing problem in estimating high frequency components. We can also see BE-WaveNet which uses features with 8kHz sampling rate can obtain magnitude spectrogram similar with WaveNet which uses the features with 16kHz sampling rate, demonstrating the effectiveness of BE-WaveNet in bandwidth extension.

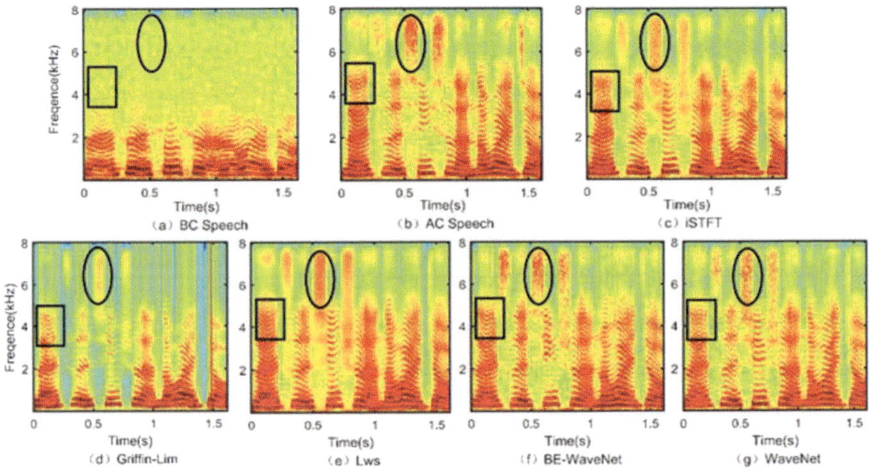

Fig. 5. Magnitude spectrograms of speech enhanced by different methods.

4 Conclusion

In order to acquire enhanced speech at high-sampling rate in BC-speech communication system and do not increase the communication cost, we use WaveNet to synthesize the speech waveform and accomplish the bandwidth extension simultaneously. In particular, a Bandwidth Extension WaveNet model is proposed, which can generate high-sampling rate speech from the enhanced magnitude with low sampling rate directly. Experimental results show that the speech quality is greatly improved compared with other waveform synthesis methods such as using the original phase of BC speech and the phase estimation methods like Griffin-Lim. Meanwhile, the proposed method shows good ability of bandwidth extension.

Demos of the enhanced speech by different methods are available online. Please visit the following website: https://github.com/echoaimaomao/Demos-for-BE-WaveNet.

5 Acknowledgments

This work was supported by the National Natural Science Foundation of China (Grant 6147139).

References

1. Ono H. Device for picking up bone-conducted sound in external auditory meatus and communication device using the same: U.S. Patent 5,295,193[P]. 1994-3-15.
2. Shin H, Kang G, and Fingscheidt T. Survey of speech enhancement supported by a bone conduction microphone [C]. 10th ITG Symposium of Speech Communication Symposium, 2012, Braunschweig, Germany.
3. Liu B, Tao J, Wen Z, Li Y, and Bukhari D. A novel method of artificial bandwidth extension using deep architecture [C]. 16th Annual Conference of the International Speech Communication Association (INTERSPEECH), 2015, Dresden, Germany.
4. Shahina A, Yegnanarayana B. Mapping speech spectra from throat microphone to close-speaking microphone: A neural network approach [J]. EURASIP Journal on Advances in Signal Processing, 2007: 1-10.
5. Trung P N, Unoki M, Akagi M. A study on restoration of bone conducted speech in noisy environments with lp-based model and gaussian mixture model [J]. Journal of Signal Processing, 2012: 409-417.
6. Toda T, Nakagiri M, Shikano K. Statistical voice conversion techniques for body-conducted unvoiced speech enhancement [J]. IEEE/ACM Trans. on Audio, Speech, and Language Processing, 2012: 2505-2517.
7. Vijayan K, Murty R. Comparative study of spectral mapping techniques for enhancement of throat microphone speech [C]. 20th IEEE National Conference on Communications (NCC), 2014, Kanpur, India.
8. Changyan Z, Xiongwei Z, Meng S, Jibin Y, Yibo X. A Novel Throat Microphone Speech Enhancement Framework based on Deep BLSTM Recurrent Neural Networks. 4th IEEE International Conference on Computer and Communications (ICCC), 2018, Chengdu, China.
9. Griffin D, Lim J. Signal estimation from modified short-time Fourier transform [J]. IEEE Transactions on Acoustics, Speech, and Signal Processing, 1984, 32(2): 236-243.
10. Williamson D S, Wang Y, Wang D L. Complex ratio masking for joint enhancement of magnitude and phase [C]. IEEE International Conference on Acoustics, Speech and Signal Processing (ICASSP), 2016, Shanghai, China.
11. Oord A, Dieleman S, Zen H, et al. Wavenet: A generative model for raw audio [J]. arXiv preprint, arXiv:1609.03499, 2016.

12. Mehri S, Kumar K, Gulrajani I, et al. SampleRNN: An unconditional end-to-end neural audio generation model [J]. arXiv preprint arXiv:1612.07837, 2016.
13. Sefik Emre Eskimez, Kazuhito Koishida, and Zhiyao Duan, Adversarial training for speech super-resolution[J], IEEE Journal of Selected Topics in Signal Processing, 2019, 13(2): 347-358.
14. Wang M, Wu Z, Kang S, et al. Speech Super-Resolution Using Parallel WaveNet[C]. International Symposium on Chinese Spoken Language Processing (ISCSLP), 2018, Taibei, China.
15. Shen J, Pang R, Weiss R J, et al. Natural TTS synthesis by conditioning WaveNet on mel-spectrogram predictions[C]// IEEE International Conference on Acoustics, Speech and Signal Processing (ICASSP), 2018, Calgary, Canada.
16. Chen K, Chen B, Lai J, et al. High-quality Voice Conversion Using Spectrogram-Based WaveNet Vocoder[C]// 19th Annual Conference of the International Speech Communication Association (INTERSPEECH), 2018, Graz, Austria.
17. Van den Oord A, Kalchbrenner N, Espeholt L, et al. Conditional image generation with PixelCnn decoders [C]. Advances in neural information processing systems (NIPS). 2016, Barcelona, Spain.
18. Yu F, Koltun V. Multi-scale context aggregation by dilated convolutions [J]. arXiv preprint arXiv:1511.07122, 2015.
19. Salimans T, Karpathy A, Chen X, et al. Pixelcnn++: Improving the pixelcnn with discretized logistic mixture likelihood and other modifications [J]. arXiv preprint arXiv:1701.05517, 2017.
20. Streijl R C, Winkler S, Hands D S. Mean opinion score (MOS) revisited: methods and applications, limitations and alternatives[J]. Multimedia Systems, 2016, 22(2): 213-227.
21. Taal C H, Hendriks R C, Heusdens R, et al. An algorithm for intelligibility prediction of time–frequency weighted noisy speech[J]. IEEE Transactions on Audio, Speech, and Language Processing, 2011, 19(7): 2125–2136.
22. Gray A, Markel J. Distance measures for speech processing[J]. IEEE Transactions on Acoustics, Speech, and Signal Processing, 1976, 24(5): 380-391.
23. Le Roux J, Kameoka H, Ono N, et al. Phase initialization schemes for faster spectrogram-consistency-based signal reconstruction [J]. In Proceedings of the Acoustical Society of Japan Autumn Meeting, 2010: 3-10.

Naturalness evaluation of synthetic speech based on residual learning networks

Yiqing Guo[1] and Jie Zhu[1]

School of Electronic Information and Electrical Engineering,
Shanghai Jiao Tong University, Shanghai 200240, China

Abstract. With the emergence of text-to-speech (TTS) systems, the synthetic speech have been widely applied. But the quality evaluation method of the synthetic speech is still away from an agreement. This paper presents a novel method based on residual learning networks, which was trained on the databases of Blizzard Challenge and our own corpus. The effect of the proposed architecture was compared with the LSTM networks and VGG networks. According to the results of experiments, the architecture based on the ResNet gives root mean square error (RMSE) of 0.40 and the correlation coefficient of 0.82, which is better than the other two networks.

Keywords: Text-To-Speech, speech evaluation, ResNet

1 Introduction

With the emergence of TTS systems [1], the synthetic speech have been widely applied, such as intelligent voice assistants, voice navigation and audio books. However there is still no automated well-established method to evaluate the quality of synthetic speech. The mainstream evaluation methods are subjectively, for example, mean opinion score (MOS) is widely used in the speech synthesis community [2], which rating the naturalness of speech as 1 to 5 by a variety of professional listeners. Obviously most of the subjective measures are time-consuming and laborious.

In contrast, objective evaluation methods can be operated in batches and consume little time and resources. The conventional objective measures such as root mean squared error (RMSE) and fundamental frequency (F_0) are widely applied to the rough quality evaluation of speech in the early stage. However, the results of these measures often differ greatly from that of human perception, and always require the corresponding natural speech as reference.

To overcome these problems, many novel objective evaluation methods have been proposed in telecommunications research, such as ITU-TP.563 [3] and ITU-TP.862 (PESQ) [4], as alternatives to evaluate the performance of TTS systems. These measures have played a significant role in the evaluation of call quality and are still widely used and researched. These measures have played a significant role in the evaluation of voice quality in telecommunication, which are still popular research topic and have been widely used. However, most of the them still have

H. Li et al. (eds.), *Proceedings of the 7th Conference on Sound and Music Technology (CSMT)*, Lecture Notes in Electrical Engineering 635,
https://doi.org/10.1007/978-981-15-2756-2_2

poor correlation with human perception. Thus, TTS systems seldom use them as criterion for evaluating the naturalness of synthetic speech.

The quality of the speech evaluation measure is dependent on the difference between the predicted results and MOS. Root mean square error (RMSE) and correlation coefficient (ρ) between the predicted score and the ground truth (MOS) have been applied as criteria of naturalness evaluation systems. As the machine learning technology tending to be mature, researchers have applied it to this field and presented the methods generalized quality score from database of human auditory perception judgement. In 2016, Yoshimura T. et al. [5] presented a CNN+LR model to automatically regress the naturalness score of the synthesized speech, which consist of a two-layer convolutional neural network (CNN) and a linear regression network (LR). The assessment system proposed gets the RMSE of 0.68 and the ρ of 0.57 on Blizzard Challenge English database.

The LSTM networks is the latest recurrent neural networks, which have been widely used in speech processing. For instance, Meng Tang el al. proposed a two-layer LSTM+LR model in 2018 [6], which performed better than two-layer CNN+LR model and got RMSE of 0.39 and ρ of 0.70 on Blizzard Challenge Mandarin database.

In 2015, Kaiming He et al. confirmed their hypothesis that residual learning is easier to optimize [7]. The ResNet is stacked up to 152 layers in the paper and perform better than previous model. The research conducted a series of comparative experiments about deep neural networks (DNNs) and ResNets, and confirmed the hypothesis that residual learning is easier than original mapping learning. In this research, the residual learning has been used in the quality evaluation of the synthetic speech, and the experiments proved that the architecture is more suitable than the networks based on LSTM.

The remainder of the paper is organized as follows: Section 2 gives information of the database we used and the feature extraction. Section 3 and 4 introduce the architecture we proposed using residual modules and the detail of experiments. The conclusions and discussion are presented in Section 5.

2 Methods

2.1 Database

One of the databases used in this research is the mandarin database of the Blizzard Challenge from 2008 to 2010 [8,13,14], which provides synthetic and natural utterances labeled with 5-point scale MOS that suits our purpose perfectly. Every year, 10 to 20 participants submit thousands of synthetic speech generated from their text-to-speech systems, which then are evaluated by a large amount of professional listeners from several angles, and the results are available. To cover all pronunciations and balance the distribution of data, we also build new database and collect totally 2600 labeled utterances in our experiments. However, this combined dataset is not so big for residual learning networks. We will get better results if there are larger datasets for training.

Among these databases, the utterances labeled more than 4 point are thought to be good enough and the natural speech achieves an average score of 4.5 point. the utterances labeled as 3 point always have incoherent pronunciations between words. And the 1- to 2-point utterances often have some serious problems, such as strange rhythm, unintelligibility and incoherence, that make listeners feel uncomfortable.

2.2 Features Extraction

Pitch, formant, mel-frequency cepstrum coefficient (MFCC) and so on are common acoustic features used in audio processing, which compress the audio into a set of coefficients reflecting certain changes in the frequency domain. Compared with time series, acoustic features have less redundancy, which can make the system pay more attention to the helpful information.

MFCC represents the short-time power spectrum of speech, which is the DCT coefficients of the Mel-frequency cepstrum [9]. Mel-frequency is related to human ear's critical frequency bandwidth, and the relationship between Mel-frequency and original frequency is approximated in Eq.(1).

$$Mel(f) = 2595 \times \lg(1 + \frac{f}{700}) \tag{1}$$

Therefore, we choose MFCC as the input features of the speech quality evaluation systems [10]. The feature matrix extracted in this paper includes 13 dimensions MFCC and logarithmic fundamental frequency ($logF_0$) with first and second derivatives. The first 14 coefficients only represent the static feature of each frame of the speech, but the first and second derivatives are able to reflect the power changes and connections between frames. The heat map of input feature is shown as Fig.1

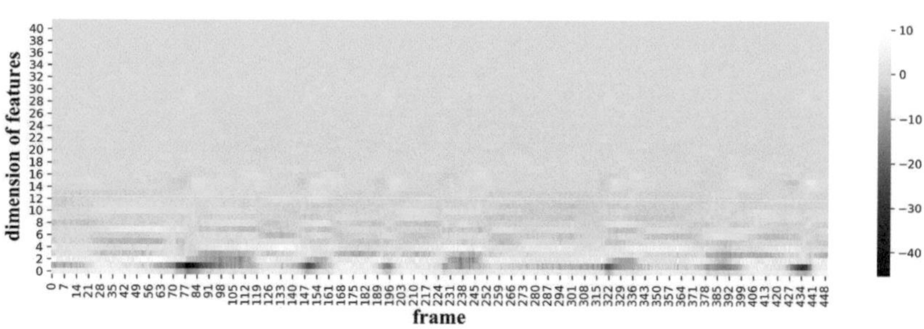

Fig. 1. The heat map of input feature: 42 dimensions coefficients extracted from each frame

Since different corpus were used in the experiments, normalization was applied to reduce the impact of acoustic difference between corpus. Therefore, the feature maps have been normalized as:

$$x' = \frac{x - m}{std} \tag{2}$$

Here x represents the feature maps, m is the mean of x, std is the variance of x and the result x' is the input of the evaluation system.

2.3 The ResNet Based Speech Evaluation System

The ResNet (Residual Neural Network) was proposed for image classification [7] and won the championship in ImageNet Large Scale Visual Recognition Competition (ILSVRC). Residual learning framework is used in their model, which is hypothesized that residual mapping is easier to learn than the original mapping, and the results of experiments have confirmed this hypothesis. Inspired by the VGG nets [11], the convolutional layers in the ResNet have very small (3×3) convolution filters, which can pay more attention to details. Up to now, the philosophy of the ResNet is widely used in a variety of tasks in different fields. Instead of training all the layers directly to fit a underlying mapping $H(x)$, ResNet hypothesizes that the stacked layers are able to learn the residual functions defined as Eq.(3) by a shortcuts connection between the block of some convolutional layers as shown in Fig.2.

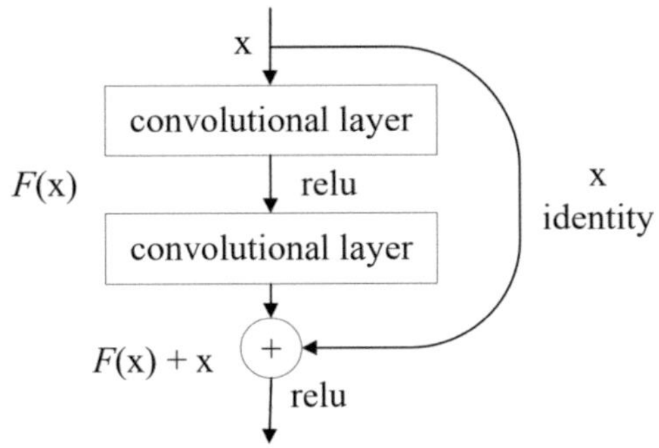

Fig. 2. A Block of Residual Network

$$F(x) := H(x) - x \tag{3}$$

The function of block which is viewed as the unit of ResNet defined as Eq. (4). Due to the shortcut connections the ResNet uses, the problems of gradient disappearance or gradient explosion in backpropagation are greatly improved, so the ResNet can be designed substantially deeper than previous networks.

$$y = F(x, \{W_i\}) + x \tag{4}$$

The time complexity of the algorithm is usually represented by the function O, which is used to qualitatively describe the running time of the algorithm. The time complexity of a CNN network is estimated as $O(\sum_{l=1}^{D} C_{l-1} \times K_l^2 \times W_l \times H_l \times C_l)$ [12]. Where D is the number of convolutional layers, C_{l-1} and C_l respectively represent the number of input and output channel of lth layer, K_l is the size of convolutional filters and W_l and H_l indicate the width and the height of the output feature, respectively. On the other hand, the time complexity of a LSTM network is $O(\sum_{l=1}^{D} w \times timestep)$ where w is the number of weights and $timestep$ is the input length.

The time complexity of complex models is difficult to calculate, but since the time spent by the algorithm is proportional to the number of the statements' executions, this paper compared the time complecity of several models by comparing the time consuming when training in the following experiments.

Fig. 3 presents the architecture proposed in our experiments, where the Residual Block is shown in Fig.4. The system takes as input MFCC feature maps of the utterances and predicts the quality scores, which are on a scale from 1 to 5 supposed to be as close to the MOS as possible. The network relies on the ResNet-18 model (weights in a total of 18 layers can be learned), which is stacked by three main parts. The first part consists of a convolutional layer with 3 × 3 filters and a max pooling layer with 3 × 3 filters. The second part is stacked by 8 residual blocks, each of which consists of two convolutional layers with 3 × 3 filters and a shortcut connection between the input and output of them. For preserving the time complexity, the number of filters is doubled when operate downsampling to the feature. The last part consists of an average pooling layer and two fully-connected layers.

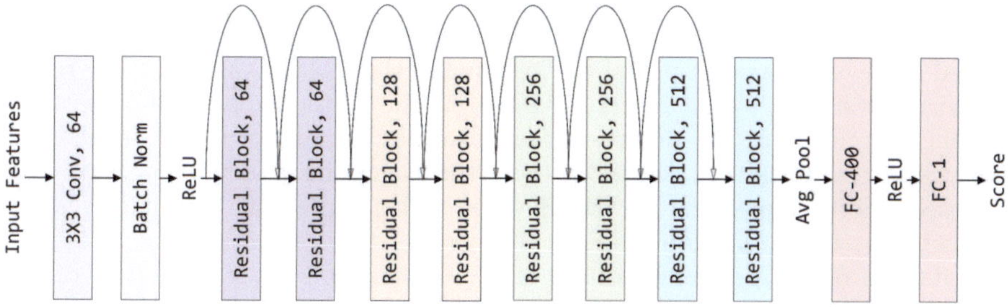

Fig. 3. Architecture of synthetic speech evaluation system based on ResNet-18

The Mean Square Error (MSE) shown as Eq.(5) was used as the loss function when training the model.

$$MSE = \frac{1}{m}\sum_{i=1}^{m}(x_i - \hat{x}_i)^2 \tag{5}$$

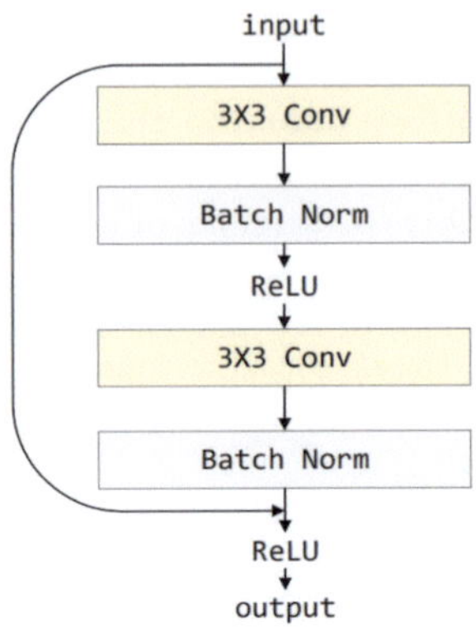

Fig. 4. The architecture of residual block

The RMSE and the correlation coefficient (ρ) were applied to objectively measure the performance of the evaluation systems in these experiments as it do in previous work [5] [6], whose expression are shown as Eq.(6) and Eq.(7). Among them, the RMSE was also used as the loss function when training model.

$$RMSE = \sqrt{\frac{1}{m}\sum_{i=1}^{m}(x_i - \hat{x}_i)^2} \tag{6}$$

$$\rho = \frac{\sum_i (x_i - \bar{x})(y_i - \bar{y})}{\sqrt{\sum_i (x_i - \bar{x})^2 \sum_i (y_i - \bar{y})^2}} \tag{7}$$

Eq.(6) describes the RMSE between the ground truth (MOS) x and the predicted score of evaluation model $\hat{x}$. The smaller the RMSE, the closer the result of scoring system is to the expected MOS, and the better the performance of the scoring system is. Eq.(7) describes the Spearman's rank correlation coefficient between x and y, here $\bar{x}$ and $\bar{y}$ represent the mean of x and y, respectively. In this scenario, x represents the MOS value and y represents the quality score predicted by the evaluation systems. The greater the correlation, the closer the distribution of the predicted results of the evaluation system is to that of the MOS, and the better the quality of the evaluation system is. Therefore, only when taking both of these two objective criteria into consideration in the same time, can we estimate the quality of the evaluation system more accurate.

3 Experiments

A total of 2600 utterances was collected as dataset including 1100 samples from the three-year Blizzard Challenge (2008-2010) [8, 13, 14] and 1800 samples from our own database. Since multiple databases have been involved, the training set and the testing set were separated randomly. And finally there are 2300 utterances as training set and 600 utterances as testing set. All the data was downsampled to 16 kHz for feature extraction.

The efficiency of the LSTM+LR network has been verified in previous work. In order to prove the validity of our architecture, two-layer LSTM+LR network and ResNet-based network were compared in this experiment, and the results of RMSE and ρ are shown in Tab.1

Table 1. The Results of Objective Measures in Different Architecture

	RMSE	ρ	Time Consuming (second)
LSTM+LR	0.52	0.64	461
ResNet-18+LR	0.40	0.82	1394

According to Tab. 1, the ResNet+LR network has smaller RMSE and higher ρ than that of the LSTM+LR network when evaluate the naturalness of the synthetic speech. Fig.5 shows the scatter plots of MOS and predicted results: LSTM+LR model and ResNet+LR model. As shown in Fig. 5, the predicted results of ResNet+LR model are more near to the ground truth. The results indicate that the ResNet can perform very well in the speech evaluation task, and even better than LSTM.

Factors including layer number, kernel size and learning method influence the testing results. In order to figure out what is the main factors lead to the results above, we adjusted the network and operated some comparative experiments. As to the layer number and kernel size, multiple convolutional layers with smaller convolution filters should be useful than less layers with larger kernels. Because the receptive field of multiple layers with smaller filters can as large as that of shallow network with larger filters. In addition, when the layer number increase, the complexity of network is increase, and the mappings networks can reflect become complex. In the beginning, the training and testing accuracy will be improved. However, as the architecture deeper to some degree, the network will be over-fitting.

Since the two-layer CNN with 3×3 filters and the one-layer CNN with 5×5 filters have the same size of receptive field, we designed a shallow ResNet as contrast to figure out the influence of deep CNN with very small convolution filters, whose convolutional layer number is halved but the size of convolution filters is 5×5 instead of 3×3. The comparison also conduct between the ResNet-18+LR model and the ResNet-34+LR model to explore the effect of the ResNet depth on accuracy. To prove the validity of the residual learning method, we created a network based on the ResNet-18+LR model but without any shortcut connections, which is similar to the VGG nets.

Tab. 2 manifests that the ResNet-18+LR model performs better than the shallow ResNet+LR model, which indicate that deep CNN networks with very small filters are more effective than shallow networks with larger filters. Com-

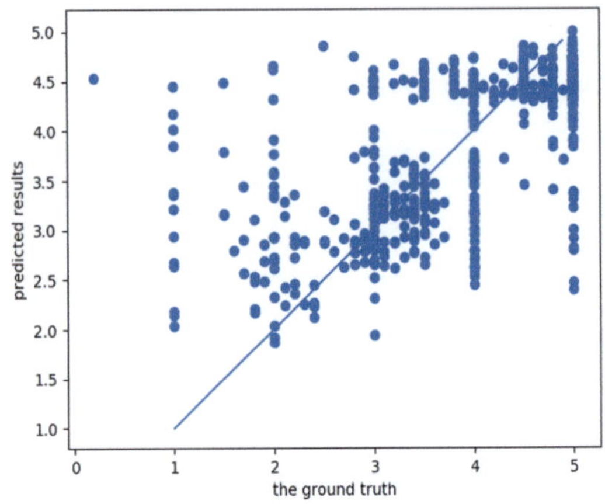

(a) LSTM+LR

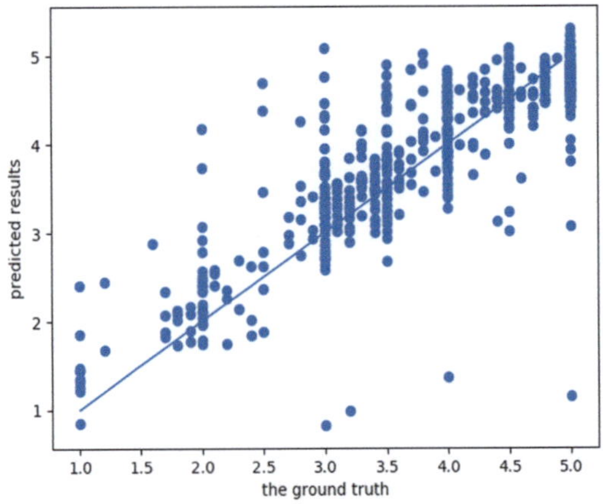

(b) ResNet-18+LR

Fig. 5. Scatter plots of MOS and predicted results: in the left and right plots, the ρ is 0.64 and 0.82 respectively pared with the ResNet-18+LR model, the VGG+LR model gets the same ρ but larger RMSE which indicates that residual learning has effect on improving the RMSE, but little effect on improving the correlation. While there is no much difference in the results of objective measures between the ResNet-18+LR model and the ResNet-34+LR model. This possibly because the number of data sets is limited, and the depth of the ResNet-18+LR model is enough for the network to converge.

Table 2. The Results of Objective Measures in Different Architecture

	RMSE	ρ	Time Consuming (second)
shallow ResNet+LR	0.48	0.71	1037
ResNet-34+LR	0.43	0.80	5017
VGG+LR	0.44	0.76	276
ResNet-18+LR	0.40	0.82	1394

According to the time consuming results of Tab.1 and Tab.2, a comparison of time complexity among three type of neural networks is obtained: ResNet $>$ LSTM $>$ VGG, which shows that the time complexity of the LSTM network is larger than that of the CNN network, and the time complexity of ResNet is much larger than the former two, and the main reason is the increased short cut between layers in the structure.

4 Conclusions And Future Work

In this paper, a novel network based on the ResNet was proposed for the synthetic speech evaluation and got a good performance. The innovation of our work is to evaluate the quality of TTS systems instead of MOS. Meanwhile, we figured out that the main factor of the improvement is the philosophy of multiple convolutional layers with small filters in the following experiments. And The residual learning method is also effective. These experiments have showed that the ResNet based network is suitable to process the speech. However, more experiments should be operated to improve the robustness and universality.

Recently, there are quite a few researches process the speech signals directly and were verified to be effective. In the future work, the format of input data is considered to be changed, for the networks are complex enough to reflect the mappings between input and output while the preprocessing of data compresses a variety of the original signals.

Acknowledgement

This work is supported by the National Key Research Project of China under Grant No.2017YFF0210903, the National Natural Science Foundation of China under Grant No.11433002.

References

1. Alexander Kain and Michael W. Macon. Spectral voice conversion for text-to-speech synthesis. In *IEEE International Conference on Acoustics*, Seattle, 1998.
2. Xianjun Xia. *Research on Speech Synthesis Method Combining Subjective Evaluation and Feedback.* PhD thesis, University of Science and Technology of China, Hefei, 2014.
3. International Telecommunication Union. Single-ended method for objective speech quality assessment in narrowband telephony applications. *ITU-T Recommendation P.563*, 2004.
4. A. W. Rix, J. G. Beerends, M. P. Hollier, and A. P. Hekstra. Perceptual evaluation of speech quality (pesq)-a new method for speech quality assessment of telephone networks and codecs. In *2001 IEEE International Conference on Acoustics, Speech, and Signal Processing. Proceedings (Cat. No.01CH37221)*, Salt Lake City, 2001.
5. Takenori Yoshimura, Gustav Eje Henter, Oliver Watts, Mirjam Wester, Junichi Yamagishi, and Keiichi Tokuda. A hierarchical predictor of synthetic speech naturalness using neural networks. In *INTERSPEECH*, San Francisco, 2016.
6. M. Tang and J. Zhu. Text-to-speech quality evaluation based on lstm recurrent neural networks. In *2019 International Conference on Computing, Networking and Communications (ICNC)*, Honolulu, 2019.
7. Kaiming He, Xiangyu Zhang, Shaoqing Ren, and Sun Jian. Deep residual learning for image recognition. In *IEEE Conference on Computer Vision & Pattern Recognition*, Las Vegas, 2016.
8. S. King, R. A. Clark, C. Mayo, and V. Karaiskos. The blizzard challenge 2008. In *Proceedings of the Blizzard Challenge workshop*, Australia, 2008.
9. Sirko Molau, Michael Pitz, Ralf Schluter, and Hermann Ney. Computing mel-frequency cepstral coefficients on the power spectrum. In *IEEE International Conference on Acoustics*, Salt Lake City, 2001.
10. Vibha Tiwari. Mfcc and its applications in speaker recognition. *International Journal on Emerging Technologies Issn*, 2010.
11. Karen Simonyan and Andrew Zisserman. Very deep convolutional networks for large-scale image recognition. *Computer Science*, 2014.
12. Eleni Tsironi, Pablo Barros, Cornelius Weber, and Stefan Wermter. An analysis of convolutional long-short term memory recurrent neural networks for gesture recognition. *Neurocomputing*, 2017.
13. S. King, R. A. Clark, C. Mayo, and V. Karaiskos. The blizzard challenge 2009. In *Proceedings of the Blizzard Challenge workshop*, Australia, 2009.
14. S. King, R. A. Clark, C. Mayo, and V. Karaiskos. The blizzard challenge 2010. In *Proceedings of the Blizzard Challenge workshop*, Australia, 2010.

Detection of Operation Type and Order for Digital Speech

Tingting Wu[1], Diqun Yan[1, 2][0000-0002-5241-7276], Li Xiang[1] and Rangding Wang[1]

[1] College of Information Science and Engineering, Ningbo University, Zhejiang Ningbo 315211, China
[2] Guangdong Key Laboratory of Intelligent Information Processing and Shenzhen Key Laboratory of Media Security, Shenzhen, 518060, China
yandiqun@nbu.edu.cn

Abstract. Most existing speech forensic works implicitly assume the suspected speech either has or has not been processed by a specific operation. In practice, however, the operation type performed on the target speech is usually unknown to the forensic analyst, and in most cases, multiple operations may be involved in order to conceal the forgery trace. Few works have considered these issues. In this study, we propose a universal forensic algorithm that can detect four typical speech operations: pitch shifting, noise-adding, low-pass filtering, and high-pass filtering. The motivation of the proposed algorithm is based on the observation that different operations will cause different effects on Mel-frequency cepstral coefficients (MFCC). The statistical moments of MFCC are extracted as detecting features. Additionally, cepstral mean and variance normalization (CMVN), which is a computationally efficient normalization technique, is used to eliminate the impact of channel noise. Finally, an ensembled binary classifier is used to detect the type of various operations, and multiclass classifiers are adopted to identify the order of operations. The experimental results on the TIMIT and UME-ERJ datasets show that the proposed forensic features achieve good performance on the operation type and order detection. Additionally, the results demonstrate the effectiveness of the proposed algorithm in terms of robustness against the MP3 compression attack.

Keywords: Speech Forensics, MFCC, Operation Trace, Multiclass Classification.

1 Introduction

With the wide availability of sophisticated editing tools (such as Audition, GoldWave, and Audacity), digital speech can be easily modified without leaving obvious artifacts. In some scenarios, such modification will cause serious moral and legal problems. For example, pitch shifting is a common speech-editing technique in which the original pitch of a digital voice is raised or lowered. This technique is often used in TV interviews and other similar instances where there is a need to protect the speak-

H. Li et al. (eds.), *Proceedings of the 7th Conference on Sound and Music Technology (CSMT)*, Lecture Notes in Electrical Engineering 635,
https://doi.org/10.1007/978-981-15-2756-2_3

er's privacy. However, this technique may also be used by malicious attackers in order to conceal their real identity.

Speech forensics is a main branch of multimedia security which aims to assess the authenticity of digital speech. Many speech forensic methods have been proposed for various speech operations [1, 2], such as pitch shifting [3, 4, 5], device source [6, 7, 8] and replaying [9]. However, several problems have not been carefully considered in previous works. First, most existing methods examine suspected speech for the effects of a single operation [10]. It is not reasonable in practice because the processing history of the suspected speech is usually unknown. For example, it will cause misleading result if a forensic detector for pitch shifting is applied on a resampled speech. Secondly, in real scenarios, the forger is likely to perform various operations on the target speech in order to hide their tracks more subtly. For example, after pitch shifting, one may use low-pass filtering, which has been specifically designed for the purpose of defeating forensics. Hence, it is meaningful to develop a forensic algorithm that can detect various operations.

In this work, four typical operations including pitch shifting, noise adding, and low-pass and high-pass filtering are investigated. Pitch shifting is always applied to conceal the manipulator's identity. Noise adding can be applied to replace the background scene of the target speech. Low-pass and high-pass filtering are often applied to remove traces of other tampering operations. Based on the observation that these operations may cause different effects on mel-frequency cepstral coefficients (MFCC), we propose an algorithm to detect various speech operations with the statistical moment features of MFCC. In order to suppress the influence of the speech content, cepstral mean-variance normalization (CMVN) is applied on the extracted features. TIMIT [11] and UME-ERJ [12] datasets have been used to evaluate the performance of the proposed method. The experimental results show that the proposed method can achieve a high detection rate and good robustness against MP3 compression attack.

The rest of this paper is organized as follows. In Section 2, the effects of various operations on MFCC are first analyzed. Then, the extraction and normalization of MFCC statistical moments are presented. Last, we propose the detection algorithm. In Section 3, a series of experimental results are presented. Finally, the conclusions are given in Section 4.

2 Proposed detection algorithm

2.1 Effect of various operations on spectrograms

Spectrogram is a common time-frequency analysis and visualization tool. It can facilitate dominant instantaneous frequency estimation in a signal as a function of time by calculating and displaying the Short-Time Fourier transform (STFT) of the speech. Fig. 1 and Fig. 2 show the spectrograms of a speech randomly selected from TIMIT dataset that has been processed with various operations.

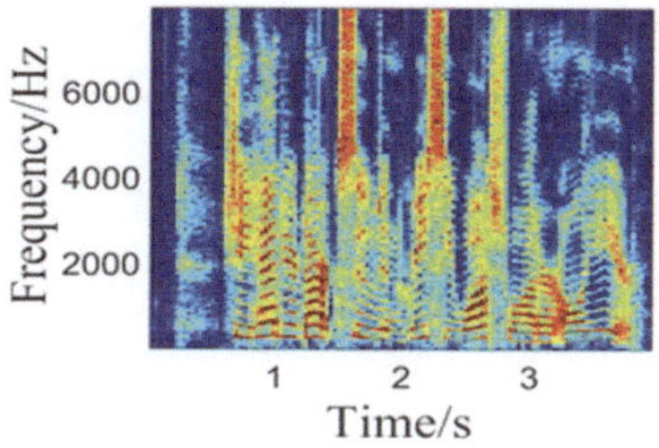

Fig. 1. Spectrogram with original speech.

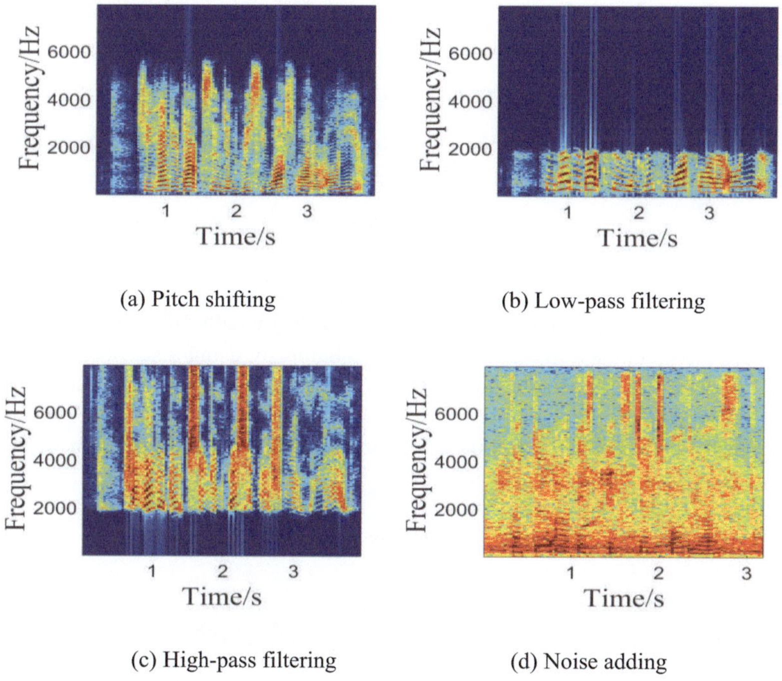

Fig. 2. Spectrogram with various operations.

From Fig. 2, it can be seen that there are obvious differences in the spectrograms with various operations. Compared to the spectrogram of the original speech (Fig. 1), the frequency bands in the pitch-shifted spectrogram are compressed. For the low-pass filtering spectrogram, most of the frequency components higher than 2000 Hz are lost. Similarly, the frequency components less than 2000 Hz are lost as a result of high-pass filtering. It should be noticed that in the pitch-shifted spectrogram, the high-frequency components are also lost, which may cause confusion between the pitch shifting speech and the low-pass filtering speech. Additionally, with noise adding, the overall energy distribution of the spectrogram is concentrated in the low frequency. It indicates that the inherent characteristics in frequency domain of the speech are difficult to preserve after these four operations, which leave more or less modification traces. By properly studying the features in the frequency domain, it is possible to distinguish the speech that has been from the original and the operated one, further, to identify the type and order of the operations.

2.2 Feature extraction

MFCC is the most widely used acoustic feature in frequency domain for speech recognition, which models the spectral energy distribution in a perceptually meaningful way. Before calculating MFCC, in order to amplify the high-frequency components, the input speech signal $x(n)$ should be first pre-emphasized using a finite impulse response (FIR) high-pass filter as follows:

$$x'(n) = x(n) - 0.95x(n-1). \tag{1}$$

Then the emphasized signal $x'(n)$ is divided into overlapped frames by multiplying each frame with a Hamming window, where N is the length of the Hamming window.

$$w(n) = 0.54 - 0.46 \cos\left(\frac{2\pi n}{N-1}\right), 0 \le n \le N-1. \tag{2}$$

Next, the spectrum is obtained by applying a Fast Fourier transform (FFT) on each windowed frame. In this work, the frame size is 30ms with 50% overlap. The spectrum is then decomposed into a number of sub-bands by using a set of triangular Mel-scale band-pass filters. Let $E(b), 0 \le b \le B$ denote the sum of the power spectrum coefficients within the bth sub-band, where B is the total number of filters (in this work, B is set to 12). MFCC can be calculated by applying discrete cosine transforms (DCT) to the logarithm of $E(b)$ as follows:

$$C(l) = \sum_{b=0}^{B-1} log_{10}\big(1 + E(b)\big) \cos\left(l\frac{\pi}{B}(b+0.5)\right), 0 \le l \le L. \tag{3}$$

where L is the length of the MFCC (L is set to in this work).

The TIMIT speech dataset, which consists of 6300 speech samples, is adopted. To evaluate the statistical difference in MFCC between the original speech and the various operation ones. For each speech, the mean values of the first 12 MFCC coefficients are calculated. The boxplots of the first four MFCC coefficients for different operations are shown in Fig. 3. It can be seen that the distributions of most feature components have significant changes after different operations. Meanwhile, there are obvious differences between the original speech and other versions in the 25th and 75th percentiles. It means that the statistics of MFCC will be affected by various operations.

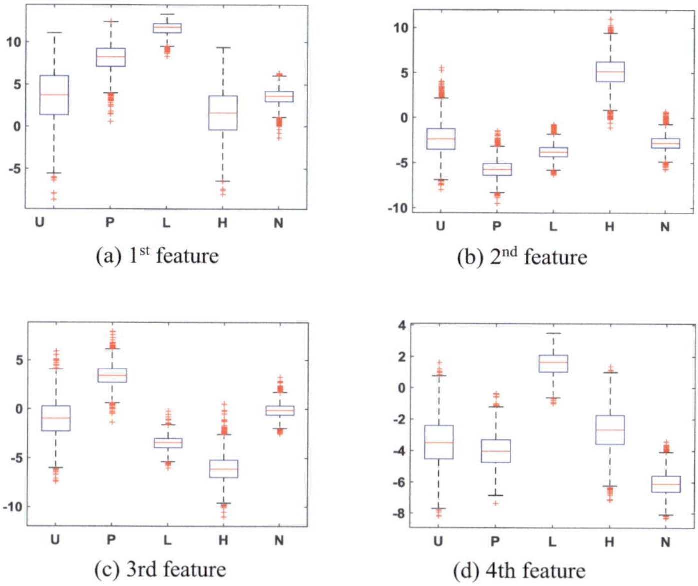

(a) 1st feature (b) 2nd feature

(c) 3rd feature (d) 4th feature

Fig. 3. Distributions of the MFCC features from an original, pitch shifting, low-pass filtering, high-pass filtering and noise-adding speech set.

Considering that the MFCC only contain static information from a given frame, the derivatives of MFCC are calculated to capture extra information about the temporal dynamics. Hence, in this work, the first 12 MFCC and their associated first- and second-order derivatives are extracted to form a 36-dimensional feature vector. Furthermore, in order to make better use of the spectral information, the statistical moments, including mean (3*12=36 dimensions) and correlation coefficients (3*66 = 198 dimensions) of the feature vector, are derived out. As a result, we have to a 234-dimensional MFCC feature vector F.

Finally, in order to suppress the effect of the channel noise, cepstral mean and variance normalization (CMVN) [13, 14], which is an efficient noise compensation technique popularly used in many speech applications, is performed on the feature vector F, where $\bar{\mu}$ and $\hat{\sigma}$ are the mean and standard variance of the feature vector F.

$$F = \frac{F - \bar{\mu}}{\hat{\sigma}} \tag{4}$$

2.3 Classification

In this work, two different forensic cases were considered. The first case is to detect various operation types rather than one specific operation in most existed works. The second case is to detect the operation order, which also involves determining which operations have been performed on the target speech.

Operation type detection. For this case, the classification strategy in [15] is adopted in this work, which uses one-versus-one strategy to train a multiclass classifier, as shown in Fig. 4. In total, we obtain $K \times (K+1)/2$ binary classifiers to detect K operations. For a given testing sample, its features are fed to these binary classifiers, and N results are obtained. In this work, K is 4 and N is 10. Finally, a majority voting strategy is adopted to determine whether the speech is original or has been processed by some specific operation.

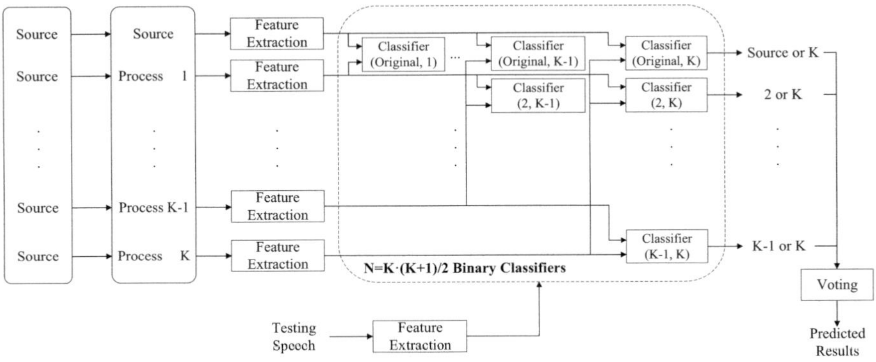

Fig. 4. Operation type detection scheme.

Operation order detection. For this case, two possible approaches could be considered. One is to extend the operation type classifier to take each operation order as a class. In this approach, if there are I operations and the maximum length of the order is J, we will have $\sum_{i=0}^{I} J^i = \frac{J^{I+1}-1}{J-1}$ different operation orders. There will be 21 orders when I is 4 and J is 2, which means that K is 21 for the operation type detection system. Therefore, a total of $N = 21 \times (21+1)/2 = 231$ binary classifiers should be trained. It is obvious that the complexity of the experiments is dramatically increased.

An alternative approach is to utilize the multiclass classifier for detecting the operation order. In this work, three typical methods, decision tree, random forest and softmax, are adopted as the classifiers.

3 Experimental results

In the experiments, the speech from TIMIT (6,300 samples) and UME-ERJ (4,040 samples) datasets are taken as natural speech. These are in WAV, 16 kHz, 16 bits, mono file format. The pitch-shifted speech is obtained by Adobe Audition (version CS6), which is a digital audio workstation from Adobe Systems. The low, high pass filtering and noise added speech samples with different parameter settings are implemented by MATLAB. The parameters of these operations are shown in Tab. 1.

Tab. 1 Type of operations and parameters used in the experiments.

Operation Type	Parameters
Pitch shifting	Modification degree: +4, +8, -4, –8
Low-pass filter	Pass-band cut-off frequency (Hz): 500, 750, 1000,1250 Stop-band cut-off frequency (Hz): 750, 1000, 1250, 1500
High-pass filter	Pass-band cut-off frequency (Hz): 1750, 2000, 2250, 2500 Stop-band cut-off frequency (Hz): 1500, 1750, 2000, 2250
Noise adding	SNR (dB): 5, 10, 15, 20

For operation type detection, there are 16 operated speech samples for each original speech. Therefore, for TIMIT dataset, we have 6,300 natural speech and 100,800 operated speech. For UME-ERJ dataset, there are 4,040 natural speech and 64,640 operated speech.

As described in Section 2.3, the proposed algorithm consists of training and testing stages. The division of the datasets for operation type detection is shown in Tab. 2. Due to the complexity of the operation order detection, only the orders with two operations are considered. Hence, for each speech, we will have 16 double-operated speech. Meanwhile, the parameter for each operation is set as follows:

1) The factor of the pitch shifting is -8.
2) The SNR of the noise adding is 5 dB.
3) The pass-band cut-off frequencies and the stop-band cut-off of the low- and high-pass filter are 500 Hz, 750Hz and 2500 Hz, 2250Hz respectively.

In following sections, the symbol '*XY*' is used to denote the double operation. For example, 'PH' means that the original speech is first shifted by -8 semitones and then processed by a high-pass filter with the cut-off frequency at 2500 Hz.

Tab. 2 Division of dataset (natural /operated).

Dataset	TIMIT		UME-EMJ	
	Training	Testing	Training	Testing
TIMIT	4000/64000	2300/36800	6300/100800	4040/64640
UME-ERJ	4040/64640	6300/100800	3200/51200	840/13400

3.1 Effect of various operations on spectrograms

Detection rate. In real scenarios, the operation parameter of the suspect speech is usually unknown to the detector. Hence, in this experiment, all operational parameters in Tab. 1 are considered. According to the description in Section 2.3, a total of $N = 5 \times 4/2 = 10$ binary classifiers are trained.

Tab. 3 Detection rate with various parameters.

Dataset	Type	Accuracy（%）							
		Parameter 1		Parameter 2		Parameter 3		Parameter 4	
TIMIT	P	+4	100	+8	99.83	-4	98.96	-8	98.96
	L	500kHz	97.39	750kHz	99.65	1000kHz	100	1250kHz	100
	H	1750kHz	100	2000kHz	100	2250kHz	100	2500kHz	100
	N	5dB	100	10dB	100	15dB	100	20dB	100
UME	P	+4	97.14	+8	99.04	-4	95.71	-8	99.52
	L	500kHz	95.71	750kHz	99.52	1000kHz	98.57	1250kHz	99.04
	H	1750kHz	100	2000kHz	100	2250kHz	100	2500kHz	98.1
	N	5dB	100	10dB	100	15dB	99.52	20dB	99.04

Tab. 4 Detection performance of inter-dataset and cross-dataset (results below 0.5% are made with *).

Training	Testing	Operation	Accuracy（%）				
			U	P	L	H	N
TIMIT	TIMIT	U	99.43	0.74	*	*	*
		P	0.57	99.26	*	*	*
		L	*	*	100	*	*
		H	*	*	*	100	*
		N	*	*	*	*	100
UME-ERJ	UME-ERJ	U	97.86	1.31	*	*	*
		P	2.14	98.21	*	*	*
		L	*	*	99.52	*	*
		H	*	*	*	99.64	*
		N	*	*	*	*	99.64
TIMIT	UME-ERJ	U	88.51	11.49	*	*	1.53
		P	7.82	83.31	*	*	*
		L	*	3.61	99.93	*	*
		H	3.32	1.55	*	99.23	*
		N	*	*	*	*	98.25
UME-ERJ	TIMIT	U	82.98	4.45	*	*	*
		P	0.81	94.73	*	*	*
		L	*	*	99.99	*	*
		H	*	*	*	99.96	*
		N	15.75	*	*	*	99.99

Tab. 3 shows the classification accuracy of the 4 parameters of each operation in TIMIT and UME datasets. The higher the accuracy, the higher the degree of modification of the original for this operation, and the proposed feature in this paper is also stronger.

Tab. 4 shows the experimental results for intra-dataset and cross-dataset experiments. In the intra-dataset experiment, the results show that with the increase of the operation strength, the detection rate increased, which is in line with the expected results. The overall detection rate was above 97%, which indicates that the proposed algorithmic features can capture traces of the operation effectively. For the cross-dataset experiment, there was a slight decrease in the detection rate, especially in the case of pitch shifting. This is mainly due to the various recording environments and the sample lengths in different datasets.

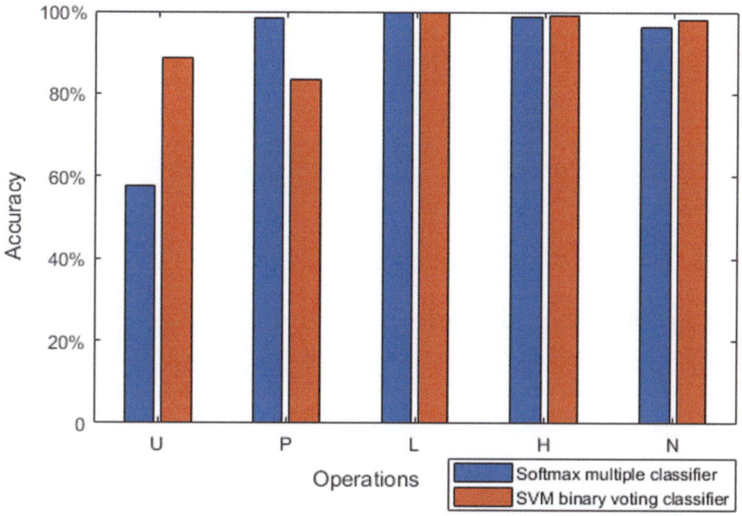

Fig. 5. Detection accuracies for multiple classifiers of Softmax and SVM binary multiple classifiers.

Comparison with Softmax classifier. Fig. 5 shows the detection performance of the Softmax multiple classifier and the SVM binary multiple classifiers. The TIMIT dataset was used for training and the UME-ERJ dataset for testing. The Softmax classifier performed better than the SVM classifier in detecting the pitch shifting and the noise addition. However, for the original speech, the Softmax classifier only achieved a low detection rate (57.5%), which means that it would cause a higher rate of false alarms. Therefore, we consider the overall performance of the SVM classifier to be better than the Softmax classifier.

Robustness of MP3 compression. In this experiment, we evaluated the robustness of the normalized MFCC statistical moment feature for MP3 compression. First, the original audio file was compressed to 128kbps and then decompressed into a WAV file. The experimental results are shown in Fig. 6. The results show that after MP3 compression, the detection performance of the algorithm was reduced by up to 3%, but the overall effect still remained above 96%. This indicates that the proposed algorithm is robust enough to handle MP3 compression.

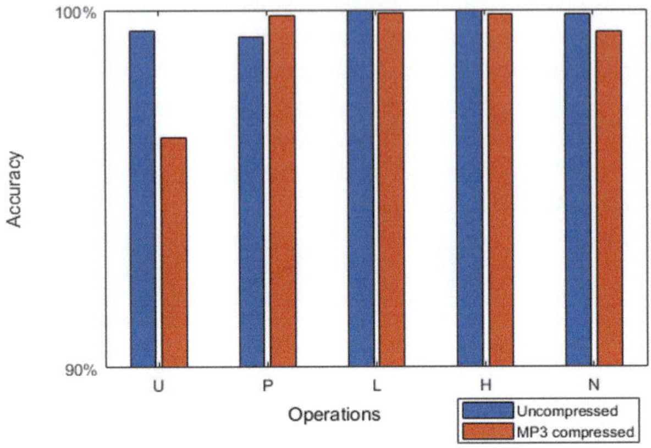

Fig. 6. Average detection accuracies for MP3 compression.

3.2 Evaluation of operation order detection

Order of operation detection with operation type detection. Due to the different methods in which the samples are processed and the intensity of the impact on the speech samples, some of the operations will mask or be masked by the traces of other operations. In this experiment, the operation type detectors mentioned in 3.1 were used to identify the order of double operations. The detection results are shown in

Tab. 5 Results of single-processed detector detecting processing chain.

	PP	PL	PH	PN	LP	LL	LH	LN
U	324	0	0	0	0	0	0	0
P	1953	0	2300	122	0	0	0	0
L	23	2300	0	0	2300	2300	2300	0
H	0	0	0	0	0	0	0	0
N	0	0	0	2178	0	0	0	2300
	HP	**HL**	**HH**	**HN**	**NP**	**NL**	**NH**	**NN**
U	0	0	0	0	0	0	0	0
P	2292	0	0	0	2300	0	0	0
L	0	2299	0	0	0	2300	0	0
H	8	1	2300	0	0	0	2300	0
N	0	0	0	2300	0	0	0	2300

Tab. 5. From the results of the last eight columns, it can be seen that if the first operation on the speech is high-pass filtering or noise adding, the operation type detectors always identify it as the second operation. This indicates that the traces of these two operations can be easily masked by other operations. All double operations with low-pass filtering are detected as low-pass filtering, whether it is the first or second operation. This means that low-pass filtering leaves a deep trace in the speech sample and is difficult to be covered by other operations.

Order detection with the multiclass classifier. Typical multiclass classifications are used to detect the operation orders. In this experiment, there are 16 different operation orders. During the training stage, the labeled training set is fed into the classifiers to train a 16-class classifier.

Tab. 6 Detection accuracies for Softmax, decision tree and random forest (%).

	PP	PL	PH	PN	LP	LL	LH	LN
Softmax	100	99.96	85.87	99.87	100	98.83	95.70	99.78
Decision Tree	99.30	96.74	66.13	95.17	99.57	77.26	70.26	90.00
Random Forest	100	99.78	67.26	99.00	100	81.61	73.09	98.48
	HP	HL	HH	HN	NP	NL	NH	NN
Softmax	100	99.35	82.91	97.35	100	100	99.78	97.26
Decision Tree	98.30	90.83	66.61	85.91	98.09	98.91	95.00	83.43
Random Forest	99.40	96.48	62.17	89.35	99.91	99.96	92.87	83.48

The experimental results are shown in Tab. 6. It can be seen that decision trees and random forest classifiers achieve similar detection results with rates higher than 90% in most cases. However, when high-pass filtering is the second operation (PH, LH, and HH), the detection accuracy is significantly decreased. The main reason is that the trace left by the high-pass filtering is relatively weak compared to other operations, such as pitch shifting or low-pass filtering. Therefore, the classifiers will ignore the trace left by the high-pass filtering and discriminate the first operation mistakenly as the second operation. For the Softmax classifier, the detection rates of most cases are better than decision trees and random forest classifiers. Similarly, there is a slight drop in the detection rate when the second operation is high-pass filtering. Additionally, the proposed classifier can achieve a 100% detection rate when pitch shifting is the second operation. From the experimental results, we can see that the proposed method can achieve good detection performance in various cases.

4 Conclusion

In this work, we propose a normalized MFCC statistical moment feature to detect various operation types and identify the operation order. Four typical operations including pitch shifting, noise adding, and low- and high-pass filtering are investigated. We have observed that these operations will leave traces in the frequency domain of the speech sample. Based on this observation, we designed an efficient feature set based on MFCC to detect the types and order of the operations.

To detect the operation type, a multiclass classifier combined with 10 binary SVM classifiers was designed. To identify the operation order, three typical multiclass classifiers were adopted. The experimental results show that the proposed method was effective in detecting the four operations and robust when handling MP3 compression. The results also indicate that low-pass filtering is less affected by other operations and that traces of high-pass filtering will be largely obscured by other operations. In further work, it is necessary to design a feature that can sense more processing operations. We will also study the parameter estimation of operation order.

5 Acknowledgments

This work was supported by the National Natural Science Foundation of China (Grant No. 61300055, U1736215, 61672302), Zhejiang Natural Science Foundation (Grant No. LZ15F020002, LY17F020010), Ningbo Natural Science Foundation (Grant No. 2017A610123).

References

1. Luo, D., Yang, R., Li, B., et al.: Detection of double compressed AMR audio using stacked autoencoder. IEEE Transactions on Information Forensics and Security 12(2), 432–444 (2017).
2. Luo, D., Yang, R., Huang, J.: Detecting double compressed AMR audio using deep learning. In: International Conference on Acoustics, Speech and Signal Processing, pp. 2669–2673. IEEE, Florence (2014).
3. Wu, H., Wang, Y., Huang, J.: Identification of electronic disguised voices. IEEE Transactions on Information Forensics and Security 9(3), 489–500 (2014).
4. Wu, H., Wang, Y., Huang, J.: Blind detection of electronic disguised voice. In: International Conference on Acoustics, Speech and Signal Processing, pp. 3013–3017. IEEE, Piscataway (2013).
5. Xu, H., Yan, D., Yang, F., et al.: Detection algorithm of electronic disguised voice based on convolutional neural network. Telecommunications Science 34(2), 46–57 (2018).
6. Luo, D., Korus, P., Huang, J.: Band energy difference for source attribution in audio forensics. IEEE Transactions on Information Forensics and Security 13(9), 2179–2189 (2018).
7. Zou, L., He, Q., Feng, X.: Cell phone verification from speech recordings using sparse representation. In: International Conference on Acoustics, Speech and Signal Processing, pp. 1787–1791. IEEE, South Brisbane (2015).

8. Qi, S., Huang, Z., Li, Y., et al.: Audio recording device identification based on deep learning. In: International Conference on Signal & Image Processing, pp. 426–431. IEEE, Beijing (2016).
9. Gałka, J., Grzywacz, M., Samborski, R.: Playback attack detection for text-dependent speaker verification over telephone channels. Speech Communication 67, 143–153 (2015).
10. Li, H., Luo, W., Qiu, X., et al.: Identification of various image operations using residual-based features. IEEE Transactions on Circuits and Systems for Video Technology 28(1), 31–45 (2018).
11. TIMIT Acoustic-Phonetic Continuous Speech Corpus, http://www.ldc.upenn.edu/Catalog/CatalogEntry.jsp?catalogId =LDC93S1, last accessed 2013.
12. Advanced Utilization of Multimedia to Promote Higher Education Reform Speech Database (UME), http://research.nii.ac.jp/src/en/UME-ERJ, last accessed 2013.
13. Prasad, N. V., Umesh, S.: Improved cepstral mean and variance normalization using Bayesian framework. In: IEEE Workshop on Automatic Speech Recognition and Understanding, pp. 156–161. IEEE, Olomouc (2014).
14. Joshi, V., Prasad, N. V., Umesh, S.: Modified mean and variance normalization: Transforming to utterance-specific estimates. Circuits, Systems, and Signal Processing 35(5), 1593–1609(2016).
15. Knerr, S.: Single-layer learning revisited: a stepwise procedure for building and training a neural network. Neurocomputing: Algorithms, Architectures and Applications 68(11), 41–50 (1990).

Part II
Music Information Retrieval

Singing Voice Detection Using Multi-Feature Deep Fusion with CNN

Xulong Zhang[1], Shengchen Li[2], Zijin Li[3], Shizhe Chen[4], Yongwei Gao[1], and Wei Li[1,5]

[1] School of Computer Science, Fudan University, Shanghai, 201203, China
[2] Institute of information photonics and optical communications, Beijing University of Posts and Telecommunications, Beijing, 100876, China
[3] Department of Musicology, China Conservatory of Music, Beijing, 100101, China
[4] Department of Music Engineering, Shanghai Conservatory of Music, Shanghai, 200031, China
[5] Shanghai Key Laboratory of Intelligent Information Processing, Fudan University, Shanghai, 201203, China

Abstract. The problem of singing voice detection is to segment a song into vocal and non-vocal parts. Commonly used methods usually train a model on a set of frame-based features and then predict the unknown frames by the model. However, the multi-dimensional features are usually concatenated together for each frame, with little consideration of spatial information. Hence, a deep fusion method of the Multi-feature dimensions with Convolution Neural Networks (CNN) is proposed. A one dimension convolution is made on feature dimensions for each frames, then the high-level features obtained can be used for a direct binary classification. The performance of the proposed method is on par with the state-of-art methods on public dataset.

Keywords: Convolution Neural Network (CNN), Multi-feature Fusion, Deep Learning, Singing Voice Detection (SVD)

1 Introduction

In the field of Music Information Retrieval (MIR), singing voice detection (SVD) is to locate the vocal portions in a piece of music, which can be seen as a useful preprocessing step for a variety of MIR tasks, such as singer identification [1], singing voice separation [2], singing voice melody transcription [3], query by humming [4], lyrics transcription etc. The main difficulty of SVD mainly comes from the extent of vocal tone diversity.

The issues of SVD are usually addressed through traditional statistical methods [5], such as Gaussian Mixture Models (GMM), neural networks and support vector machines (SVM) [6], Hidden Markov Model (HMM) [7] and etc. Eyben et al. [8] proposed the data-driven approach based on Long Short-Term Memory Recurrent Neural Networks (LSTM-RNN) for Voice Activity Detection (VAD) in speech. The main advantage of the LSTM model is the ability to

H. Li et al. (eds.), *Proceedings of the 7th Conference on Sound and Music Technology (CSMT)*, Lecture Notes in Electrical Engineering 635,
https://doi.org/10.1007/978-981-15-2756-2_4

model long range dependencies between input series. For the successful use of LSTM on numerous research areas, Lehner et al. [9] introduced the LSTM to SVD. There are up to 111 audio features used in feature representation. The use of LSTM achieves the state-of-art performance on the two publicly available datasets (Jamendo [6] and RWC [7]). Leglaive [10] added bi-directional structure on LSTM that takes the past and future temporal context into account on the presence/absence of singing voice. In Schlüter's work [11], Convolutional Neural Networks (CNN) model on Mel spectrograms is used to design the singing voice detection system. The CNN model has been demonstrated powerful to learn invariances taught by data augmentation in other fields.

Except the design of classifier in singing voice detection, another important module is feature representation. These studies [12][13][14]are based on features mostly come from speech processing field, such as Mel-Frequency Cepstral Coefficients (MFCC) and Linear Predictive Cepstral Coefficients (LPCC). However, these features may not be good enough to distinguish the singing voice from background music. Rocamora and Herrera [15] found that MFCC and their derivatives are the most appropriate features, the accuracy was around 78.5%.

Regnier and Peeters [12] presented a method to detect vocal segments within an audio track based on two specific characteristics of the singing voice, vibrato and tremolo. In [16], Lehner et al. optimized the MFCC features with manually tuned parameters, they achieved 82.36% accuracy after a temporal smoothing. In another work [17], three new features are designed by Lehner. Though the results are competent to other works, the process of feature extraction is too complicated.

Single feature cannot fully describe the audio features, and concatenation features together simply may lead incompatible between different features or highly dependent on a certain dimension of the feature sets. Therefore, we can treat each feature dimension as a one dimensional space. Due to the space relation information, CNN is used to learn invariant features, which can relieve the complexity of manual design features and make different dimensions' feature more compatible. In addition to CNN models for the fusion features, a preprocess of singing voice separation is applied to get more focus on the vocal part and a post-process of temporal smoothing to modify the obvious temporal exception.

The rest of this paper is structured as following. An overview of our SVD system is presented in section 2. Experiments and results on common benchmark datasets are then presented and discussed in section 3. Finally, some conclusions are proposed in section 4.

2 Proposed SVD system

The architecture of the proposed system uses singing voice separation (SVS) as the preprocess step to get vocal signal, then follows a traditional bag-of-frames approach: a machine learning technique (CNN) is applied on a set of features computed on successive frames of the incoming vocal signal. The output of the classifier is then further being temporal smoothed to localize musical segments that contain singing voice. The overview of our SVD system is shown in figure 2, and different building blocks are described in detail below.

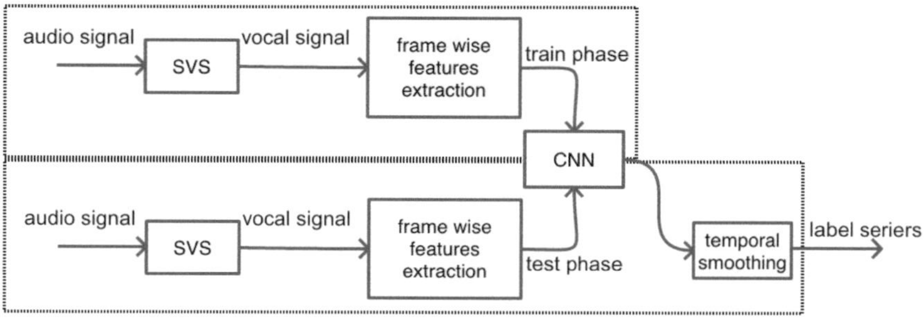

Fig. 1. The proposed SVD system overview

2.1 Singing Voice Separation

Singing voice separation (SVS) is taken as a preprocess step, which can split the mixed music signal into vocal portions and accompaniment portions. SVS method is based on the "REPET-SIM" method [18] with modifications and extensions: FFT windows are overlapped by 1/4, instead of 1/2. Non-local filtering is converted into a soft mask by Wiener filtering. This is similar in spirit to the soft-masking method used by [19], but is a bit more numerically stable in practice. Nevertheless, the split of accompaniment portions and vocal portions is hard to be completely separated due to the strong accompaniment, the separated vocal contains a little harsh noise. After SVS we save three versions of audio signal on the same dataset, then we can compare their results with the same classifier.

2.2 Feature Extraction

This section briefly describes the chosen features. There are many features proposed for the singing voice detection problem. To Among these features, MFCC [20], LPCC [21] and Chroma [22] are examined in this paper. The three features were chosen for feature fusion cause that they describe mixed audio at three different aspects. MFCC has been widely used in many speech and audio recognition tasks [15] and MFCC can represent the timbre of the audio signal. LPCC features are calculated by introducing the cepstrum coefficients (CCs) to the Linear Predicative Coding (LPC) parameters. LPCC feature reveals the nature of the produced sound which is governed by the shape of the vocal tract. The Chroma features are a well-established tool for analyzing and comparing music data [23].

The audio signal is segmented into 40ms frames with an overlapping of 20ms. FFT is calculated on each frame with Hamming window. Most features were selected for their ability to discriminate vocal with music [24].

The features were calculated on short-scale frames stated above. 26 MFCC coefficients (without any energy coefficient), 12 LPCC coefficients and 12 Chroma coefficients were extracted from each frame. Finally, the combined feature vector has 50 dimensions (MFCC-26, Chroma-12, LPCC-12).

2.3 CNN for feature deep fusion and Classification

In this paper, we present a novel data-driven method for singing voice detection based on deep fusion of features with the CNN model. The motivation behind the use of CNN is the capacity to learn compositional representations in spatial, where invariants from the original feature spatial can be learned. It not only can extract deep features, but also can use these learned feature for the binary classification directly.

The proposed networks for SVD have an input layer which matches the size of the combined acoustic feature vectors, two one-dimension Convolutional layers, two one-dimension Max pooling layers, and dense layers to flatten the input to the output layer with a single sigmoid unit.

The input of the CNN is successive frames. The multi-features were extracted and then concatenate together to build the raw feature representation of each frame. Then there are two dimensional inputs: x axis represents the temporal series of frames while the y axis represents different dimensions of features. The one-dimension Convolutional will process on x axis where the flatten layer will combine the final extracted features of different frames. Connect them to the dense layer for output.

The networks are trained as classifier to output a voicing score for every frame in the value space of 0 and 1:1 indicating singing frame, 0 indicating no singing frame. The final step in predicting the audio frame at timestamp t is to take a softmax and then do a temporal smoothing over the outputs of the sequential model. The neural network topologies are shown in figure 2.3.

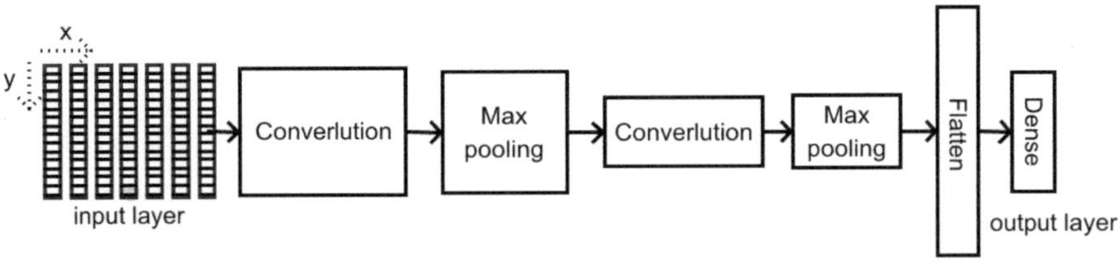

Fig. 2. The neural network topologies of the proposed CNN

2.4 Post Process

Due to the classifications vary significantly. The likelihood ratios vary wildly from frame to frame. This is in stark contrast to the out labeling data where the class labels stay the same for many successive frames. Given that the singing voice has continuity in a certain period, it is more reliable to accumulate the segment likelihood over a longer period of decision making.

In this paper, we proposed three methods for segmentation smoothing. The first one is the median filter to smooth the raw classification variable along the time dimension. The second method is to use the posterior probabilities obtained by a Hidden Markov Model of two states (vocal and non-vocal). The observation distributions are modeled by a mixture of 45 Gaussians, fitted with the Expectation Maximization algorithm. The best path of states is then deduced from the classifier output sequence with the Viterbi algorithm. The third one is the Conditional Random Field (CRF) that was used to learn the relation of the prediction and the ground truth on validation dataset.

3 Experiments and results

For comparison, we choose two public mainly used datasets for SVD. We also make common evaluation on accuracy, precision, recall and F1 measure.

We compare the performance of the difference features to determine how helpful they are and how to parameterize them, and then combine the best features. Besides, the best performance system on the two public datasets against our implemented LSTM under the same conditions (feature representation, preprocess and post-process) are also compared.

3.1 Benchmark Datasets

To our knowledge, there are two publicly available corpora with vocal activity annotations. One is Jamendo corpus and the other is RWC pop music dataset.

Jamendo Corpus [6] has a set of 93 songs, which constitute a total of about 6 hours of music. Each file was annotated manually into singing and non-singing sections by the same person to provide the ground thrust-data. As Jamendo Corpus had been split the 93 songs into 3 parts to generate train, test and valid sets. We use the same split dataset for training and testing as the compared relate work.

The RWC Popular Music Dataset [7] contains 100 pop songs, with singing voice annotations by Mauch. As this dataset commonly was used by 5-fold cross validation and use the average performance for compilation. The same split was done as the compared relate work.

3.2 Evaluation

In order to give a comprehensive view of the results, we compare model predictions with the ground truth labels to obtain true positives(TP), false positives

(FP) true negatives (TN) and false negatives (FN) over all songs in the test set. Besides, we also calculate the frame-wise accuracy, precision, recall and with F1 measure to summarize results. These metrics can be represented in below equations:

$$Accuracy = \frac{TP + TN}{TP + FP + TN + FN} \tag{1}$$

$$Precision = \frac{TP}{TP + FP} \tag{2}$$

$$Recall = \frac{TP}{TP + FN} \tag{3}$$

$$F1 - measure = \frac{2 * precision * recall}{precision + recall} \tag{4}$$

3.3 Results

In the first experiment, the mixed audio of music signal was split into vocal and accompaniment and the 13 MFCC coefficients were extracted as the audio feature to train the classifier without the post process on the prediction. We compare binary classification results on split vocal, the split accompaniment and the mix audio of music signal. The performance on the preprocessed data and mix audio is showed in figure 3.3. We use 'jamendo vocal' to denote the SVD result of vocal part after singing voice separation on Jamendo dataset. And with the 'jamendo mix' represent the result of Jamendo raw data without any preprocess of SVS. 'jamendo music' is the result of the music part after SVS of Jamendo dataset. The result labeled on RWC dataset was similar at Jamendo.

From the comparison of the classification results use different audio signal on two datasets. The use separated vocal is higher than the raw mix signal by 2% in accuracy. The accompaniment music signal is lower than the raw mix signal, so we can conclude that apply the preprocessing of singing voice separation can improve the final vocal detection performance. In the music, the accompaniment often strong and not only overlapped with the vocal in temporal it also intertwined with vocal in frequency. So do a preprocess of singing voice separation can degrade the influence by the accompaniment.

We compare the performance of different features that we choose in section 2.2. In the first experiment, three different features and their combinations (LPCC, Chroma, MFCC) are compared to classify the vocal and non-vocal segments using the deep CNN model separately. Through this experiment we want to check if the CNN can obtain more effectively information from the single features. The performance of different features on separating vocal part of RWC and jamendo dataset is shown in figure 3.3 and 3.3.

From figure 3.3 and 3.3, the Chroma feature is not fit for the vocal detection task. Although MFCC was very popular in the relate work such as in [16], in this experiment, MFCC performed normally. The LPCC has almost the best performance compared with the other two features. When the three features'

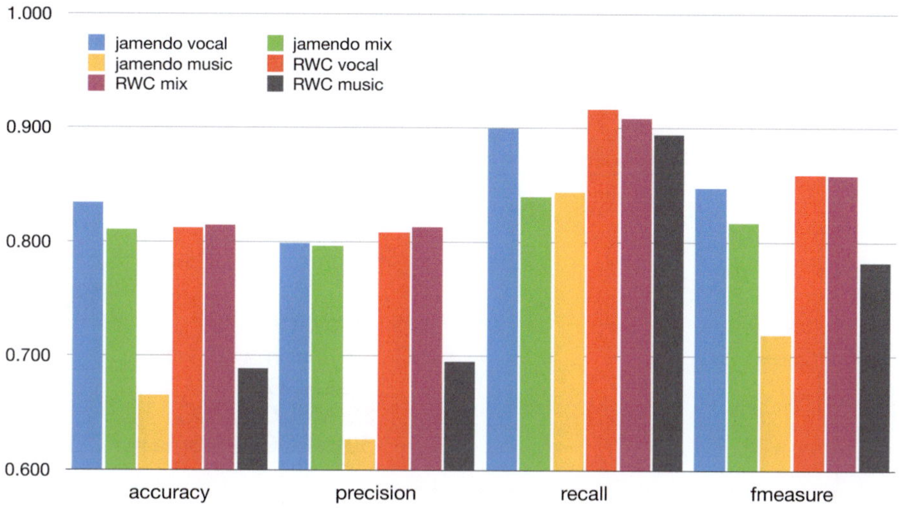

Fig. 3. Different audio data with singing voice separation as preprocess

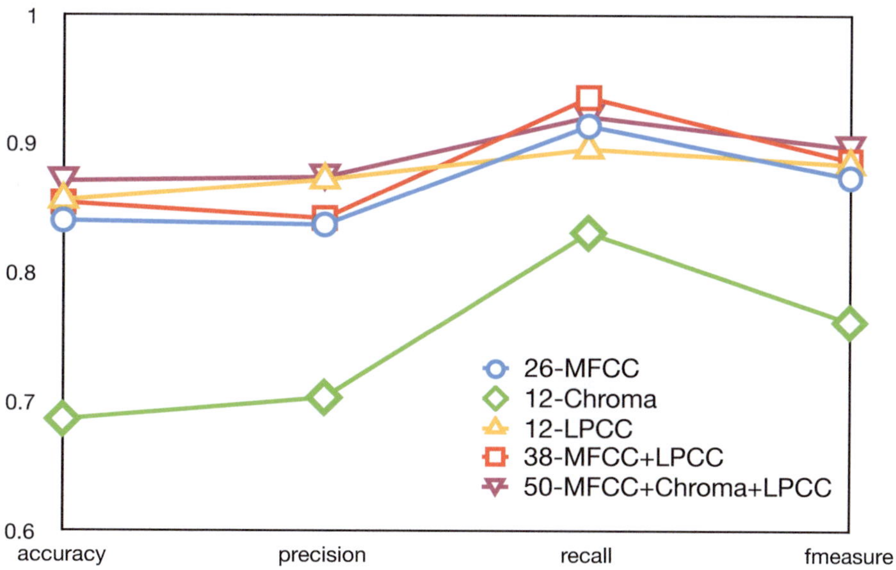

Fig. 4. Different features performance on RWC dataset

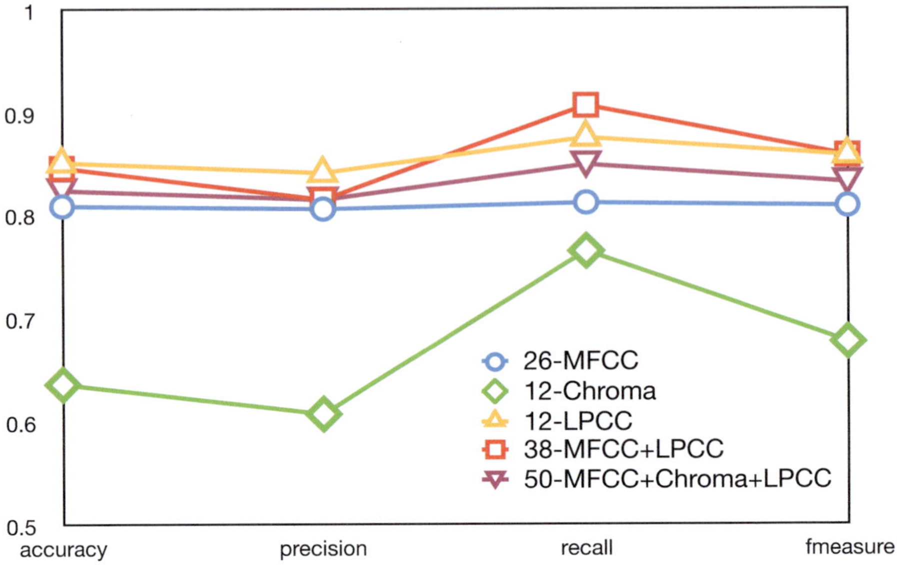

Fig. 5. Different features performance on Jamendo dataset

combination vector was chosen and feed to CNN model. CNN fusion different dimensions of the feature well. Therefore, the worst performance of Chroma feature does not affect the overall performance much.

In order to verify the effect of post-processing, the different temporal smoothing method were compared in the post process described in section 2.4. Firstly, median filtering was used for temporal smoothing and the number is decided by experiment on validating dataset. In the second, the HMM was used on the predict probability that HMM based method not need the fixed length frames window. The total series probability was used to train the HMM and then the prediction of the final segments boundary can be got. Due to HMM was used as an unsupervised model, it just uses the information of the classifier's prediction probability. Given a probability sequence, a 2-state HMM was trained on it, and at the end of the training process, run the Viterbi algorithm on the sequence to get the most likely state associated with each input vector. The segment boundary was found by HMM, and then the frames from each segment vote for the final label. The third is CRF with nearly the same as HMM, but the main difference is that CRF is a supervised model. The validation dataset was used to get the prediction and the ground truth, then train the CRF model. The trained CRF model is used for predicting the smoothing label series.

From the comparison result show in figure 3.3, the post process is necessary for singing voice detection. Compare with the blue line one without using post processing which labeled as 'Without postprocess', the performance can be improved by 4% by HMM. The performance of smoothing process of median fil-

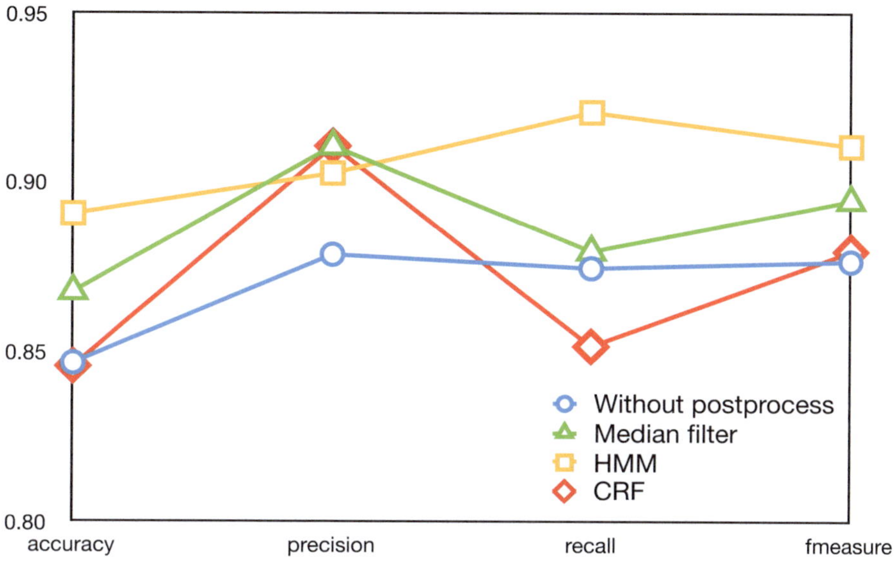

Fig. 6. With different temporal smoothing on RWC dataset

tering and CRF are both weaker than the case of HMM. Median filtering smooths the sequence in a fixed window and it leads the original boundary disappears and generates new fixed length segments. So use median filtering will produce more false positives, so the recall is getting smaller. As to the CRF model, it needs data sequence split into parts. Although there is no need to set each part of a fixed length in the training phase. When use the trained model need set the length, so there also has a boundary problem. If the length either too small or too long will lead to inconsistencies in training and testing.

On the public datasets we used, there are several works achieve the state of art on the task of singing voice detection. Finally, the proposed systems are compared with Ramona [6], Schlüter [11], Lehner-1 [17], Lehner-2 [16], Lehner-3 [9], Leglaive [10] on Jamendo corpus. And compare with Mauch [7], Schlüter [11], Lehner-1[17], Lehner-2[16], Lehner-3[9] for RWC pop dataset. The comparison results are presented in table 3.3 and table 3.3.

Our model is called proposed CNN in Table 1. A LSTM network is also used as a baseline system for comparison with the same preprocess and post process, which are called imLSTM.

Table 3.3. shows the comparison results on Jamendo dataset. We implement imLSTM and CNN by Keras and run it on GPU to get our results, while the other 6 results are reported in the related report on the public dataset Jamendo. The results demonstrate that on this dataset, Leglaive (uses the BLSTM-RNN) still keep the state-of-art best performance. For imLSTM with the combined three features, the F1 measure value is 0.796 which is lower than the Ramona's

Table 1. Proposed SVD System Compared with Others on Jamendo Corpus

	Accuracy	Precision	Recall	F1
Ramona	0.822	-	-	0.831
Schlüter	0.923	-	0.903	-
Lehner-1	0.882	0.880	0.862	0.871
Lehner-2	0.848	-	-	0.846
Lehner-3	0.894	0.898	0.906	0.902
Leglaive	0.915	0.895	0.926	0.910
imLSTM	0.795	0.897	0.716	0.796
Proposed CNN	0.859	0.917	0.796	0.853

SVM. But with proposed CNN model, the F1 measure gets an improvement with 5 percent. Although it has not yet reached the best performance, CNN is valid to fusion different feature dimensions to a compatible way and the result is better than the LSTM under the same conditions.

Table 2. Proposed SVD System Compared with Others on RWC Pop Dataset

	Accuracy	Precision	Recall	F1
Schlüter	0.927	-	0.935	-
Mauch	0.872	0.887	0.921	0.904
Lehner-1	0.875	0.875	0.926	0.900
Lehner-2	0.868	0.879	0.906	0.892
Lehner-3	0.923	0.938	0.934	0.936
imLSTM	0.868	0.902	0.887	0.894
Proposed CNN	0.890	0.911	0.912	0.911

Table 3.3. shows the comparison results on RWC pop dataset. On this dataset, the-state-of-art best results are kept by Lehner-3 (uses LSTM and well-design feature sets). For imLSTM with the combined three features, the F1 measure value is 0.894, it is on par with other 5 methods except the best results. Finally, the proposed CNN gets F1 measure value of 0.911, only latter than the-state-of-art best result. Compared to these two datasets, we can find that the performance on RWC is better than Jamendo. It may be because of data labeling difference between these two dataset. Jamendo corpus was labeled by one person, while the RWC pop dataset was labeled by a team. There may be some errors for manual labeling.

4 Conclusions

In this paper, a novel SVD system based on CNN with the fusion of multi-feature dimensions was proposed. In the SVD system, vocal is separated out of the mix audio signal, and CNN was used to fusion the different features dimension of the same frame. With a post processing of temporal smoothing, the performance of the proposed SVD based on CNN can be on par with the state-of-art performance on public dataset.

For future works, we will investigate the performance of CNN in more detail, such as analyzing the context learning behavior using time-frequency domain features or modulation spectrum features. Besides, we will learn compositional representations in spatial and temporal domain. Make combination of LSTM and CNN to the SVD. Furthermore, semi-supervised and active learning could be used to efficiently adapt the generic models presented in this paper to other tasks such as singer identification.

5 Acknowledgement

This work is supported by NSFC 61671156.

References

1. Kim Y, Whitman. B. Singer identification in popular music recordings using voice-coding features[C]. ISMIR. Paris, France, 2002.
2. Vembu S, Baumann. S. Separation of vocals from polyphonic audio recordings[C].ISMIR. London, UK, 2005.
3. Salamon e a J. Melody extraction from polyphonic music signals: Approaches, applications, and challenges[J]. IEEE Signal Processing Magazine. 2014.
4. Hsu e a C.-L. A tandem algorithm for singing pitch extraction and voice separationfrom music accompaniment[J]. IEEE Transactions on Audio, Speech, and LanguageProcessing. 2012.
5. Sonnleitner e a R. A simple and effective spectral feature for speech detection in mixedaudio signals[C]. DAFx'12. York, UK, 2012.
6. Ramona G R M., David B. Vocal detection in music with support vector machines[C].ICASSP. Las Vegas, NV, USA, 2008.
7. Mauch e a M. Timbre and melody features for the recognition of vocal activity andinstrumental solos in polyphonic music[C]. ISMIR. Miami, Florida, USA, 2011.
8. Eyben e a F. Real-life voice activity detection with lstm recurrent neural networks andan application to hollywood movies[C]. ICASSP. Vancouver, BC, Canada, 2013.
9. Lehner G W B., Bock S. A low-latency, real-time-capable singing voice detectionmethod with lstm recurrent neural networks[C]. EUSIPCO. Nice, France, 2015.
10. Leglaive R H S., Badeau R. Singing voice detection with deep recurrent neural networks[C]. ICASSP. South Brisbane, Queensland, Australia, 2015.
11. Schlüter J, Grill T. Exploring data augmentation for improved singing voice detectionwith neural networks[C]. ISMIR. Malaga, Spain, 2015.
12. Regnier L, Peeters G. Singing voice detection in music tracks using direct voice vibrato detection[C]. ICASSP. Taipei, Taiwan, 2009.

13. Pikrakis e a A. Unsupervised singing voice detection using dictionary learning[C].EUSIPCO. Budapest, Hungary, 2016.
14. Li X F W., Xue M. Reducing manual labeling in singing voice detection: An activelearning approach[C]. ICME. Seattle, WA, USA, 2016.
15. Rocamora M, Herrera P. Comparing audio descriptors for singing voice detection inmusic audio files[C]. Brazilian Symposium on Computer Music. San Pablo, Brazil,2007.
16. Lehner R S B., Widmer G. Towards light-weight, real-time-capable singing voice de-tection[C]. ISMIR. Curitiba, PR, Brazil, 2013.
17. Lehner G W B., Sonnleitner R. On the reduction of false positives in singing voicedetection[C]. ICASSP. Florence, Italy, 2014.
18. Rafii Z, Pardo. B. Music/voice separation using the similarity matrix[C]. ISMIR.Porto, Portugal, 2012.
19. FitzGerald D. Vocal separation using nearest neighbours and median filtering[C]. ISSC.Maynooth, Ireland, 2012.
20. You Y C W S.D., Peng S H. Comparative study of singing voice detection methods[J].Multimedia tools and applications. 2016.
21. Gupta H, Gupta D. Lpc and lpcc method of feature extraction in speech recognition system[C]. 6th International Conference-Cloud System and Big Data Engineering(Confluence). Noida, India, 2016.
22. Ellis D, Poliner G. Identifying cover songs' with chroma features and dynamic programming beat tracking[C]. ICASSP. Honolulu, Hawaii, USA, 2007.
23. Muller S E M., Kreuzer S. Making chroma features more robust to timbre changes[C].ICASSP. Taipei, Taiwan, 2009.
24. Richard M R G., Essid S. Combined supervised and unsupervised approaches forautomatic segmentation of radiophonic audio streams[C]. ICASSP. Honolulu, Hawaii,USA, 2007.

A Multi-task Learning Approach for Melody Extraction

Zhengyu Cao[1], Xiangyi Feng[2], and Wei Li[12]

[1] School of Computer Science, Fudan University, Shanghai, China, 201203
[2] Shanghai Key Laboratory of Intelligent Information Processing, Fudan University, Shanghai, China, 201203

Abstract. Melody extraction aims to produce a sequence of frequency values corresponding to the pitch of the dominant melody from a musical recording, comprising a large variety of algorithms spanning a wide range of techniques. In this paper, a novel DNN-LSTM based architecture is proposed for melody extraction. Melody extraction is regard as a composition of pitch estimation and voicing detection. This paper present a multi-task learning approach so as to perform the two tasks simultaneously, which proves to help the model obtain higher accuracy and better generalization ability. Experiments on public datasets show that the proposed model is capable of modeling temporal dependencies, and have a comparable result to the state-of-the-art methods.

Keywords: Melody extraction, Multi-task learning

1 Introduction

Melody is one of the most important factors in music. In recent years, the extraction of melody has received substantial attention from music information retrieval (MIR) communities, comprising a large variety of algorithms spanning a wide range of techniques, such as query-by-humming (QBH) [1], cover song identification [2], music transcription [3] and music structuring [4], etc. According to Poliner et al. [5], melody is the single (monophonic) pitch sequence that a listener might reproduce when asked to hum or whistle a polyphonic piece of music, and that a listener would recognize as being the essence of that music when heard in comparison. In general, the most dominant pitch sequence is considered as the melody.

Despite the variety of proposed approaches, melody extraction remains highly challenging. Apart from overlapping harmonics and high degrees of polyphony in polyphonic music, one of the complexities and challenges of this task is to determine when the melody is present and when it is not. In the context of melody extraction in polyphonic music, there are primarily two major tasks in the proposed system, i.e., 1) pitch estimation (to estimate the pitch of the melody) and 2) voicing detection (to identify the presence or absence of the melody). Almost all existed melody extraction methods focus on the former task to date. To solve the latter one, generally a subsequent voicing detection step is included.

H. Li et al. (eds.), *Proceedings of the 7th Conference on Sound and Music Technology (CSMT)*, Lecture Notes in Electrical Engineering 635,
https://doi.org/10.1007/978-981-15-2756-2_5

The most common approach is to use static or dynamic thresholds on energy or salience, often involving careful parameter tuning in order to reach peak performance [6–8]. Salamon & Gómez [9] defined a set of contour characteristics and devised rules to distinguish between melodic and non-melodic contours by exploiting their distributions. Bittner et al. [10] perform voicing detection by setting a threshold on the contour probabilities produced by the discriminative model.

Furthermore, since collections of royalty-free music recordings that can be shared for research purposes are relatively scarce, and on the other hand the annotation process for melody is difficult and tedious, there are few public datasets for melody extraction, making this task even more difficult. Most existing deep learning approaches, focusing on pitch estimation, only make use of nearly half samples of the datasets, for which melody is present. Park & Yoo [11] quantize melody pitches such that each pitch corresponds to a pitch class, and assign region without melody to a special pitch class. In this way, voicing detection is performed by comparing the special pitch class with the pitch class, however, such kind of strategy is likely to suffer data imbalance theoretically in the case when not enough training samples are available.

In this paper, a novel yet simple DNN-LSTM based architecture is proposed for melody extraction, which is capable of modeling temporal dependencies and has comparable result to the state-of-the-art methods. Differing from most existing methods, we regard melody extraction as a composition of two subtasks, i.e., pitch estimation and voicing detection. To extract melody for polyphonic music, we present a multi-task learning approach so as to perform the two tasks simultaneously. By jointly training them, we show that the model can obtain higher accuracy and better generalization ability. For reproducibility, we will share the source code at https://github.com/beantowel/Lab_MelExt.

2 Related Works

2.1 Deep Learning for Melody Extraction

More recently, deep learning has shown great advantages and potential, and has been successfully adopted in many areas. To our best knowledge, however, as for melody extraction, there is little work exploring such kind of techniques in the literatures. In these methods, the melody extraction problem is viewed as a classification problem by categorizing melody pitches into a finite set of pitch labels. Kum et al. [6] present a classification-based approach for melody extraction on vocal segments using multi-column deep neural networks, each of the networks is trained to predict a pitch label with different pitch resolutions and their outputs are combined to infer the final melody contour. Bittner et al. [12] use a fully convolutional neural network to learn salience representations from harmonic constant-Q transform representations of music signal and estimate the melody line from those representations by choosing the frequency bin with the maximum salience for each frame. Park & Yoo [11] propose a long short-term memory recurrent neural network (LSTM-RNN) for extracting melody, which is considered capable of representing the dynamic variations in melody pitch sequence. This paper treats the melody extraction as a frame-level classification problem similar to the above works, by adopting a multi-task approach, the proposed method skipped hard-coded post processing steps widely used in related works to detect voicing frames, replacing it with a deep-learning counterpart.

2.2 Multi-task Learning

In multi-task learning, more than one loss function, corresponding to the 'tasks', is optimized simultaneously. It has been used across lots of applications of machine learning, from natural language processing [13] and speech recognition [14] to computer vision [15]. Multi-task learning can be viewed as a form of inductive transfer, which help improve a model by introducing an inductive bias and thereby cause a model to prefer some hypotheses that explain more than one task. In general, there are two common approaches to perform multi-task learning in the context of deep learning, hard parameter sharing and soft parameter sharing. In hard parameter sharing, some hidden layers are shared among all tasks, while some layers including output layers are task-specific. In soft parameter sharing, different models are trained for different tasks with some constraints introduced, such as the regularized distance between the parameter of the models. For more details, an overview of multi-task learning in deep neural networks has been presented by Ruder [16]. The proposed model was based on hard parameter sharing to get a representation useful for both pitch estimation and voicing detection tasks.

3 Proposed Method

3.1 Architecture

As aforementioned, melody extraction consists of two primary subtasks, pitch estimation and voicing detection. In this paper, we believe that the two tasks are closely related in some way. Therefore, by constructing a joint two-task model, not only can we obtain predictions for both tasks at once, but also they likely benefit each other so that a much more explicit and generalized model is possible.

The proposed method extract melody in a frame-wise manner, by predicting instant frequency every 23 milliseconds. For pitch estimation, we treat it as a classification task, as done by Kum et al. [6], Bittner et al. [12] , Park & Yoo [11], Ellis & Poliner [17]. A pitch range of nearly 5 octaves from 55Hz to 1.76kHz are taken into account, i.e., from A1 to A6, so the model can capture most of the pitches in vocal melody and keep the data distribution of pitch classes roughly balanced. For voicing melody frames, the algorithm is expected to return a frequency value matching the ground truth, which is considered correct if it is within 50 cents (i.e., half a semitone) to the ground truth [9]. Instead of the 50-cents interval used by the evaluation metric, we adopt smaller pitch class interval in the model output, since we focus on vocal melody whose pitch curves are smooth and continuous due to natural singing styles such as pitch transition patterns or vibrato [18]. An 1/9 semitone resolution is used, which results in 540 pitch classes in total for the pitch estimation output. Given a ground truth instant frequency f Hz, its corresponding pitch class $Clz(f)$ is calculated as:

$$Clz(f) = \lfloor N_{pitchClass} \cdot log_{\frac{f_{high}}{f_{low}}}(\frac{f}{f_{low}}) \rfloor \tag{1}$$

where $N_{pitchClass}$ is the total number of pitch classes, and $[f_{low}, f_{high})$ corresponds to the range of pitch frequency, mapping to $[0 : N_{pitchClass})$, the range of pitch classes. Frequencies out of range were rectified to maximum or minimum values.

The voicing detection determines whether the melody is present or not at some point as a binary classification task. Notice that there are more classification classes for pitch estimation than those for voicing detection, the former task is considered relatively complicated. As a result, pitch estimation is taken as a main task, with voicing detection taken as an auxiliary task. The architecture of multi-task DNN-RNN model and a simpler DNN-only variant is illustrated in Figure 1, based on hard parameter sharing. The model takes several frames from the STFT of the input audio as inputting feature vector, and outputs two labels for each timestamp, corresponding to the predictions of pitch estimation and voicing detection respectively. Then the two labels are merged so as to obtain the melody prediction at the last stage. The lower layers, i.e., layers of deep neural network (DNN), are shared across both tasks, while top layers, i.e., long short term memory networks (LSTMs), are task-specific.

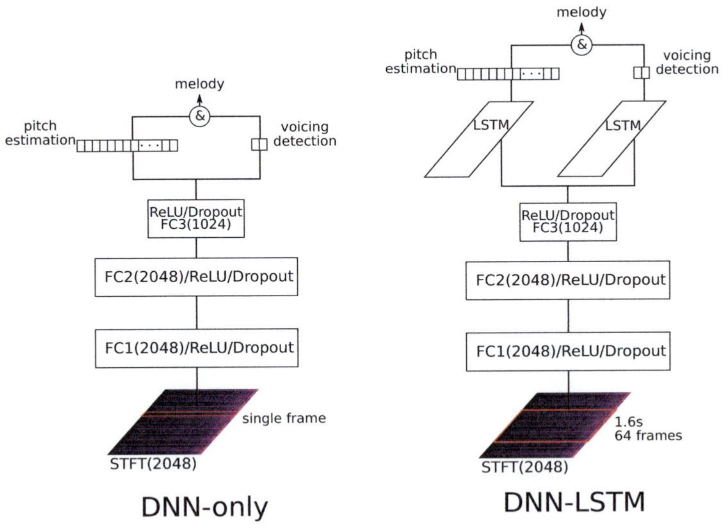

Fig. 1. The architecture of DNN-only and DNN-LSTM model for melody extraction.

In the model, DNN plays a role for representation learning where three hidden layers are used, with 2048, 2048 and 1024 units and ReLUs [19] are adopted as the nonlinear function. While the DNN takes only one frame from the STFT spectrogram for one-off prediction independently, thereby no contextual information is taken into account, which is demonstrated to be important for not only pitch estimation but also voicing detection. To capture temporal dependencies, LSTMs take the output of the last hidden layer of DNN as input, i.e., the representation learnt by DNN, which is supposed to benefit all relevant tasks. Proverbially, Long Short Term Memory networks (LSTMs) [20] are a special kind of RNN, which is capable of learning long-term dependencies and connecting previous information to the present task, and they work tremendously well on a large variety of problems. For each task (pitch estimation and voicing detection), a 2-layer bi-LSTM with 64 units in the cell is trained, thus the model will have an input vector of shape (64, 2048), capturing a $\frac{63\times512+2048}{22050} \approx 1.6$s long audio clip with the sampling rate set to 22050Hz, the STFT window set to 2048 frames long and the hop size set to 512 frames long. The last two fully connected layers will make predictions for pitch estimation and voice detection tasks respectively. To integrate the two predicted results for melody extraction, supposing we have predicted the labels for the two tasks, $label_{pe} = Clz(f) \in [0 : N_{pitchClass})$ for pitch estimation, where function Clz from Equation 1 mapped the frequency to pitch classes and $label_{vd} \in \{0, 1\}$ for voicing detection, where 0 indicates absence of melody and 1 indicates presence of melody, then melody extracted is given by:

$$instantFreq = \begin{cases} Clz^{-1}(label_{pe}), & \text{if } label_{vd} = 1. \\ 0, & \text{if } label_{vd} = 0. \end{cases} \tag{2}$$

3.2 Multi-task Loss

There are two output layers in the proposed system, one for pitch estimation and the other for voicing detection. The output layers produce probabilities over the pitch/voice classes, thus we adopt cross entropy losses by converting ground truth melody label to one-hot vector for each frame. Furthermore, a blurring technique is used as described by Bittner [12], which add a smoothing term to the one-hot vector for pitch estimation making it obey a rectified discrete Gaussian distribution centered at target pitch class with a quarter-semitone deviation σ. For non-voice frames, the target vector for pitch estimation is set to have a small value ϵ on all pitch classes. These method would help in training since the model have a high semitone resolution, nearby frequencies still make a good prediction for the melody extraction task, and they better utilise non-voice data for pitch estimation task. To balance the number of terms in loss functions involved in $loss_{pe}$, loss for pitch estimation, and $loss_{vd}$, loss for voicing detection, we multiply $loss_{vd}$ by a factor of semitone resolution N_{se}. Also, l_2 norm penalty with $\lambda = 10^{-4}$ is included to alleviate over-fitting. Multi-task loss used in the proposed model is shown in (6), where y is the target vector for pitch estimation, y_j corresponds to the component for jth pitch class, x_{pe} is the output for pitch estimation, z is the one-hot target vector for voice detection, x_{vd} is the output for voice detection and parameter $\alpha \in [0,1]$ is an auxiliary weight, which controls how cross entropy loss of the auxiliary task (voicing detection) plays a part in the global loss, i.e., 0 means only pitch estimation is concerned while 1 means the two tasks make equivalent effect. Since the losses were computed for each frame, we need to sum it up in the end.

$$y_j = \begin{cases} e^{-\frac{(j-label_{pe})}{2\sigma^2}}, & \text{if } label_{vd} \geq 0 \text{ and } |j - label_{pe}| \leq \frac{N_{se}}{2}. \\ \epsilon, & \text{if } label_{vd} < 0. \\ 0, & \text{otherwise.} \end{cases} \tag{3}$$

$$\mathbf{loss_{pe}} = -\sum_j \mathbf{y_j} \cdot log(Softmax(\mathbf{x_{pe}})_j) \tag{4}$$

$$\mathbf{loss_{vd}} = -N_{se} \sum_{j=0}^{1} \mathbf{z_j} \cdot log(Softmax(\mathbf{x_{vd}})_j) \tag{5}$$

$$L = sum\{\mathbf{loss_{pe}}\} + \alpha \cdot sum\{\mathbf{loss_{vd}}\} + \lambda \cdot \|\mathbf{w}\|_2 \tag{6}$$

3.3 Training

Since in the experiments training DNN and RNN jointly does not result in good convergence, a strategy of training DNN and RNN separately is adopted. In order to learn a representation with DNN that is helpful for both pitch estimation and voicing detection, we first train a simplified version of the model by removing LSTM layers and directly feed the DNN output to the last fully connected layers, using the same loss as Equation 6. Then we train the full model setting the parameters fixed except for those of LSTMs, i.e., using a pre-trained DNN when training LSTM layers. This can be viewed as a transfer learning technique.

All networks are trained using gradient descent with learning rate 10^{-4} and 70% dropout [21] for all hidden layers to alleviate over-fitting. The model with is implemented using Pytorch [22], an open-source software library for machine learning, and run on a computer with a single GPU with 8GB memory.

3.4 Inferencing

The proposed model takes multiple frames as input to capture temporal dependencies, original STFT spectrogram were segmented into groups of frames in a similar pattern used when slicing samples to frames in STFT. Frames in different groups will overlap in time axis, thus we use a vote mechanism when generating final outputs. As depicted in Figure 2, at timestamp t, if the count of non-voice predictions is greater equal then that of voice, the output will be non-voice, otherwise the median frequency of the predictions will be used as the pitch estimation output.

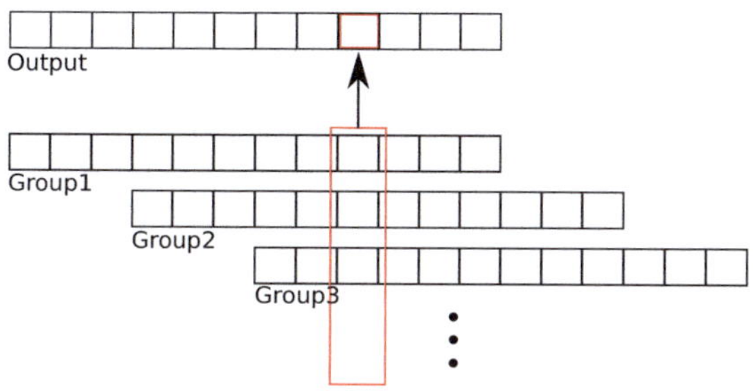

Fig. 2. Overlapping predictions vote for final output.

4 Experiments

4.1 Datasets

Training and Validation Datasets This work use the MedleyDB [23] dataset in the training phase, which contains 108 songs with MELODY2 annotation in total. It covers a wide range music styles including classical, jazz, rock and pop. Among them, 47 songs is instrumental music, the other 61 songs is vocal. Since instrumental music has clearly different characteristics from vocal music and the task of distinguishing the instrumental melody from the accompaniment is more complicated comparing to the vocal music, this work focus on melody extraction for vocal music, where singing voice is regarded as a main source of the melody, and thereby voicing detection here de facto is semantically equivalent to singing voice detection (SVD). The 61 vocal songs from MedleyDB is divided into two sets by a random split, 90% of its songs makes up the training set, the other 10% forms the validation set.

Test Datasets ADC2004[3] and MIREX-05[3] datasets are used for test. The datasets consists of 20 and 13 audio clips across different music styles respectively. Among them, all the 12 and 9 clips carrying vocal melody are used in the experiments.

4.2 Metrics

Following Poliner et al. [5], several evaluation metrics for melody extraction, including voicing recall rate (VR), voicing false alarm rate (VFA), raw pitch accuracy (RPA), raw chroma accuracy (RCA) and overall accuracy (OA) are used as a measure of performance. They are defined as follows.

1. VR: The proportion that a frame which is truly voiced is labeled as voiced.
2. VFA: The proportion that a frame which is not actually voiced is labeled as voiced.

[3]https://labrosa.ee.columbia.edu/projects/melody/

3. RPA: The proportion of voiced frames for which pitches are considered correct.
4. RCA: The proportion of voiced frames for which chroma are considered correct (octave errors are ignored).
5. OA: The proportion of all frames correctly estimated by the algorithm, including both pitch and voicing.

4.3 Effects of Auxiliary Weight for Multi-task Learning

As mentioned before, the auxiliary weight α in this work controls how much the loss of the auxiliary task (voicing detection) contributes to the total loss of the model during training. To test the effects of multi-task approach and figure out an optimal value of auxiliary weight, the experiment verify pitch accuracy (PA)[4], voicing accuracy (VA)[5] and overall accuracy with α range from 0.1 to 0.9 with step size 0.2. In addition, these metrics are computed on the simplified DNN-only model, i.e., no temporal information and LSTM layers is taken into account for faster training. All hyper-parameters except the auxiliary weight remain unchanged. The models with highest OA (i.e., top model for melody extraction) on validation dataset within 10 training epochs is evaluated for each auxiliary weight $\alpha > 0$, since we expect a model for melody extraction ultimately and their PA, VA as well as OA are presented in Table 1. Note that when α is set to 0, only the pitch estimation task is concerned, thus VA and OA do not make sense any more in this case. A DNN model for voicing detection only is also evaluated, by removing the first term in the loss function as Equation 6 while setting α to 1, denoted in the table as N/A. Similarly, in this case, PA and OA should be ignored. Therefore, when α is 0 or N/A, it is actually training a single-task model for either pitch estimation or voicing detection, and when α is larger than 0, a joint two-task model is used.

Table 1. Top models for melody extraction with different auxiliary weights and their accuracies, values labeled with * should be ignored.

α	PA	VA	OA
0	**0.7398**	0.6285*	0.4726*
0.1	0.6288	0.7516	0.7105
0.3	0.6308	0.7529	0.7118
0.5	0.6311	0.7554	**0.7139**
0.7	0.6443	**0.7606**	0.712
0.9	0.6432	0.7597	0.7128
N/A	0.0012*	0.7424	0.2973*

[4]mean raw pitch accuracy of voicing frames
[5]mean voicing accuracy of all frames

In Table 1, the model reaches the best performance in terms of OA when the auxiliary weight is set to 0.5. Comparing to single-task learning (α is 0 or N/A), multi-task learning (α ¿ 0) shows a relatively clear performance boost for voicing detection, though there are no significant differences in the performance for different settings of α when α is greater than 0. As is expected, the integration with the auxiliary task (voicing detection) can benefit the melody extraction task to some extent. Consequently, α is set to 0.5 and the model with the highest OA is used for the following comparison.

4.4 Comparison with State-of-the-Art Methods

The proposed model is trained on the MedleyDB [23] dataset and compared to 4 classical and state-of-the-art methods on 2 test datasets described in Section 4.1. Five evaluation metrics for melody extraction mentioned before are computed using mir_eval [24]. Among the compared algorithms, Salamon [9][6] is based on salience function while Bittner [12][7], Hsieh [25][8] and Lu [26][9] is based on deep neural networks, using a data-driven approach. Proposed model is denoted as MultiDNN and MultiDNN_RNN, the former is the simplified model without LSTM layers, and the latter is the full model.

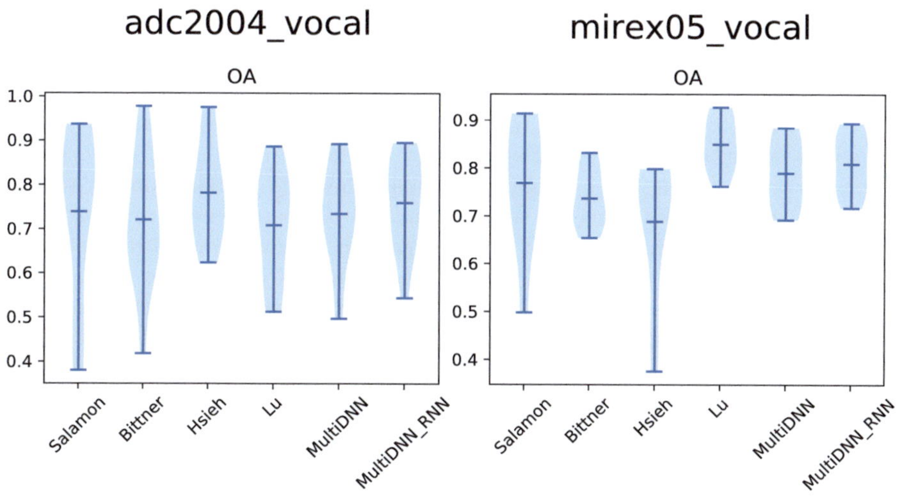

Fig. 3. OA distribution of vocal melody extraction results.

[6]https://github.com/justinsalamon/audio_to_midi_melodia
[7]https://github.com/rabitt/ismir2017-deepsalience
[8]https://github.com/bill317996/Melody-extraction-with-melodic-segnet
[9]https://github.com/s603122001/Vocal-Melody-Extraction

Table 2. Comparison of vocal melody extraction results.

Adc2004(vocal)					
Method	VR	VFA	RPA	RCA	OA
Salamon	81.5%	12.1%	77.8%	80.8%	73.9%
Bittner	83.0%	36.9%	81.3%	**85.8%**	72.1%
Hsieh	96.5%	45.9%	**83.9%**	85.3%	**78.1%**
Lu	73.9%	2.4%	67.0%	69.3%	70.9%
MultiDNN	77.8%	11.8%	71.4%	74.8%	73.4%
MultiDNN_RNN	83.7%	17.8%	74.9%	77.8%	75.9%
MIREX-05(vocal)					
Method	VR	VFA	RPA	RCA	OA
Salamon	87.0%	22.8%	80.4%	81.7%	76.8%
Bittner	79.1%	24.5%	80.1%	82.3%	73.6%
Hsieh	94.8%	42.2%	75.1%	76.4%	68.9%
Lu	87.4%	7.9%	**80.9%**	**82.5%**	**84.9%**
MultiDNN	75.6%	7.5%	70.4%	71.9%	78.9%
MultiDNN_RNN	79.0%	7.3%	73.2%	74.3%	80.9%

Figure 3 shows the min, max, mean of vocal melody extraction results based on songs(not frames), the results is slightly different from that of the original papers. For producing comparable evaluations, the code available online is used to calculate the results using the same evaluation program instead of citing original papers. As shown in Table 2, on average, the proposed model reaches high performance on par with the top-notch algorithms. Note that for the models evaluated using deep learning methods, Lu [26] and Hsieh [25] trained on MedleyDB and MIR-1K, while Bittner [12] trained on a subset of MedleyDB. Thus evaluation results on Adc2004 and MIREX-05 datasets are comparable. Among the five metrics, RPA and RCA measure the performance of the pitch estimation task while VA and VFA measure the performance of the voicing detection task. For another thing, OA combines the performance of both tasks to give an overall performance score for the system. As a consequence, comparison merely focus on the metric OA in this part. The proposed model is second only to Hsieh for Adc2004 in terms of OA, as for MIREX-05, the proposed model is second only to Lu. However, top methods on one test set was outperformed by the proposed model on the other test set, so the model has better generalization ability. It is worth mentioning that the architecture in this work is fairly simple and during training, only the MedleyDB training set is used without any additional datasets or data augmentation, which proves to be capable of improving performance in Kum [6].

5 Conclusions and Discussions

In this paper, we propose a novel DNN-LSTM based architecture for melody extraction. Notice that melody extraction can be decomposed into pitch estimation and voicing detection, we adopt the idea of multi-task learning and present a multi-task learning approach for this task, so as to perform pitch estimation and voicing detection simultaneously. Experiments show that by joint training the two tasks, the model can obtain higher accuracy and has better generalization ability. In addition, we show how the auxiliary weight takes effect in this work. Although the architecture is fairly simple and no more additional data or data augmentation is involved, the proposed model reach comparable performance to the state of the art. By expanding train dataset or increasing the complexity of the architecture, the proposed approach can be further improved.

6 Acknowledgements

This work was supported by the National Natural Science Foundation of China (Grant 61671156).

References

1. Justin Salamon, Joan Serrà, and Emilia Gómez. Tonal representations for music retrieval: From version identification to query-by-humming. *International Journal of Multimedia Information Retrieval*, 2(1):45–58, March 2013.
2. Joan Serrà, Emilia Gómez, and Perfecto Herrera. Audio Cover Song Identification and Similarity: Background, Approaches, Evaluation, and Beyond. In Janusz Kacprzyk, Zbigniew W. Raś, and Alicja A. Wieczorkowska, editors, *Advances in Music Information Retrieval*, volume 274, pages 307–332. Springer Berlin Heidelberg, Berlin, Heidelberg, 2010.
3. Emmanouil Benetos, Simon Dixon, Dimitrios Giannoulis, Holger Kirchhoff, and Anssi Klapuri. Automatic music transcription: Challenges and future directions. *Journal of Intelligent Information Systems*, 41(3):407–434, December 2013.
4. Geoffroy Peeters. Sequence Representation of Music Structure Using Higher-Order Similarity Matrix and Maximum-Likelihood Approach. In *ISMIR*, pages 35–40, Vienna, Austria, 2007.
5. Graham E. Poliner, Daniel P. W. Ellis, Andreas F. Ehmann, Emilia Gomez, Sebastian Streich, and Beesuan Ong. Melody Transcription From Music Audio: Approaches and Evaluation. *IEEE Transactions on Audio, Speech and Language Processing*, 15(4):1247–1256, May 2007.
6. Sangeun Kum, Changheun Oh, and Juhan Nam. Melody Extraction on Vocal Segments Using Multi-Column Deep Neural Networks. In *ISMIR*, pages 819–825, New York City, USA, 2016.
7. Benoit Fuentes, Antoine Liutkus, Roland Badeau, and Gael Richard. Probabilistic model for main melody extraction using Constant-Q transform. In *2012 IEEE International Conference on Acoustics, Speech and Signal Processing (ICASSP)*, pages 5357–5360, Kyoto, Japan, March 2012. IEEE.

8. Karin Dressler. Towards computational auditory scene analysis: Melody extraction from polyphonic music. *Proc. CMMR*, pages 319–334, 2012.
9. Justin Salamon and Emilia Gómez. Melody extraction from polyphonic music signals using pitch contour characteristics. *IEEE Transactions on Audio, Speech, and Language Processing*, 20(6):1759–1770, 2012.
10. Rachel M. Bittner, Justin Salamon, Slim Essid, and Juan Pablo Bello. Melody Extraction by Contour Classification. In *ISMIR*, pages 500–506, Malaga, Spain, 2015.
11. Hyunsin Park and Chang D. Yoo. Melody extraction and detection through LSTM-RNN with harmonic sum loss. In *2017 IEEE International Conference on Acoustics, Speech and Signal Processing (ICASSP)*, pages 2766–2770, New Orleans, LA, March 2017. IEEE.
12. Rachel M. Bittner, Brian McFee, Justin Salamon, Peter Li, and Juan Pablo Bello. Deep Salience Representations for F0 Estimation in Polyphonic Music. In *ISMIR*, pages 63–70, Suzhou, China, 2017.
13. Ronan Collobert and Jason Weston. A unified architecture for natural language processing: Deep neural networks with multitask learning. In *Machine Learning, Proceedings of the Twenty-Fifth International Conference (ICML 2008), June 5-9, 2008*, pages 160–167, Helsinki, Finland, 2008.
14. Li Deng, Geoffrey Hinton, and Brian Kingsbury. New types of deep neural network learning for speech recognition and related applications: An overview. In *2013 IEEE International Conference on Acoustics, Speech and Signal Processing*, pages 8599–8603, Vancouver, Canada, 2013. IEEE.
15. Ross Girshick. Fast R-CNN. In *Proceedings of the IEEE International Conference on Computer Vision*, pages 1440–1448, Santiago, Chile, 2015.
16. Sebastian Ruder. An Overview of Multi-Task Learning in Deep Neural Networks. *arXiv:1706.05098 [cs, stat]*, June 2017.
17. Daniel P. W. Ellis and Graham E. Poliner. Classification-based melody transcription. *Machine Learning*, 65(2):439–456, December 2006.
18. Sangeun Kum and Juhan Nam. Joint Detection and Classification of Singing Voice Melody Using Convolutional Recurrent Neural Networks. *Applied Sciences*, 9(7):1324, March 2019.
19. Vinod Nair and Geoffrey E. Hinton. Rectified linear units improve restricted boltzmann machines. In *Proceedings of the 27th International Conference on Machine Learning (ICML-10)*, pages 807–814, Haifa, Israel, 2010.
20. Sepp Hochreiter and Jürgen Schmidhuber. Long short-term memory. *Neural computation*, 9(8):1735–1780, 1997.
21. Nitish Srivastava, Geoffrey Hinton, Alex Krizhevsky, Ilya Sutskever, and Ruslan Salakhutdinov. Dropout: A simple way to prevent neural networks from overfitting. *The Journal of Machine Learning Research*, 15(1):1929–1958, 2014.
22. Adam Paszke, Sam Gross, Soumith Chintala, Gregory Chanan, Edward Yang, Zachary DeVito, Zeming Lin, Alban Desmaison, Luca Antiga, and Adam Lerer. Automatic differentiation in PyTorch. In *31st Conference on Neural Information Processing Systems (NIPS 2017)*, page 4, Long Beach, CA, USA, 2017.
23. Rachel M. Bittner, Justin Salamon, Mike Tierney, Matthias Mauch, Chris Cannam, and Juan Pablo Bello. MedleyDB: A Multitrack Dataset for Annotation-Intensive MIR Research. In *ISMIR*, volume 14, pages 155–160, Taipei, Taiwan, 2014.
24. Colin Raffel, Brian McFee, Eric J. Humphrey, Justin Salamon, Oriol Nieto, Dawen Liang, Daniel PW Ellis, and C. Colin Raffel. Mir_eval: A transparent implementation of common MIR metrics. In *In Proceedings of the 15th International Society*

for Music Information Retrieval Conference, ISMIR, Taipei, Taiwan, 2014. Citeseer.
25. Tsung-Han Hsieh, Li Su, and Yi-Hsuan Yang. A Streamlined Encoder/Decoder Architecture for Melody Extraction. *arXiv:1810.12947 [cs, eess]*, October 2018.
26. Wei Tsung Lu and Li Su. Vocal Melody Extraction with Semantic Segmentation and Audio-symbolic Domain Transfer Learning. In *ISMIR*, pages 521–528, Paris, France, 2018.

A post-processing of onset detection based on verification with neural network

Mingtai Lin[1] and Yin Feng[1]

Department of Artificial Intelligence, Xiamen university, 361000

Abstract. Onset detection is the primary task of music transcription that aims to find the start time of each note, which directly associated with the beats perception in the auditory system. Researchers attempted to find a data representation of universal onset function. However, the onset detection would not generalize to all cases. For example, onset detection in solo singing has a lower performance than solo playing the instrument in MIREX challenge every year. This paper presents a post-processing step to singing onset detection that solely reduces false detected onsets. In the post-processing step, the system checks the onsets picked from local maximums of onset function, and uses the neural network model to discern onset or non-onset feature rather than consider a complicated onset function. The performance of the network has a close relationship to the onset detection. In the public dataset about the research of singing transcription, the pipeline with post-processing presents a higher performance than the standard and novelty method, when it was focused on the onsets, that it reduces false alarms from feature methods. It can provide further supports for the research of singing transcription when the data-driven approach provided an effective method to eliminate spurious peaks, which can be the state-of-art of singing onset detection.

Keywords: Onset detection, Deep learning

1 Introduction

People can track beats unconsciously when they heard a piece of music. Moreover, the musician can record music score, singing transcription, which aims to record a note sequence from an input of singing audio. The framework of transcription [1] primarily consists of onset detection [2] and pitch tracking [3], which notate the sound of the singing that was previously un-notated. Automatic singing transcription system converts the audio file into note sequence in the music score, with the pitch tracking module compute the fundamental frequency (F0) of monophonic audio, and the module of note segmentation splits the F0 sequences into segments and assigns note positions to the music score. In other words, onset detection finds the position to split the pitch sequence. The reliable approaches to pitch detection includes YIN [4], PYIN [5], SWIPE [6], CREPE [?], that can generate F0 sequence. Methods of split pitches have two main scenarios, according to the threshold or feature, for example, thresholds of

H. Li et al. (eds.), *Proceedings of the 7th Conference on Sound and Music Technology (CSMT)*, Lecture Notes in Electrical Engineering 635,
https://doi.org/10.1007/978-981-15-2756-2_6

the pitch-time curve [7], and principle feature of the spectrogram of harmonic structure [8]. The purpose is to detect the onset and offset time of a note, threshold [7] or feature [8] methods are useful to singing transcription. Also, there are many onset detection algorithms that the transcription system can focus on, then achieve good singing transcription.

1.1 Common onset detection

The research proposes an universal method to find onsets in music signal when playing an instrument or singing a song, that consists of onset function and peak-picking algorithm. To deal with the special case that singing transcription [1, 9, 7, 8], the specific research for characters of solo singing can be more effective than universal onset detection. Onsets often occurs with a sudden increase in energy, spectral, phase according to the textbook of music information retrieval [10], or rather that onset of a note may be determined by locating time positions where the amplitude envelope of signal starts to increase [2]. As a result, more refined methods to compute a onset function by analyzing the spectral content of signal [2, 11, 12]. Researchers attempt to find an onset function according to energy, spectral, phase or complex-domain, to reflect the vital feature of onset or non-onset that higher or lower value to represent the probability of onset. The novelty function is a prevalent topic that based on above feature, and many researchers have put forward some innovative onset function, for example, high-frequency content [13], the flux of spectral or complex domain function [14], and design a pick-picking algorithm to pick onsets in onset function. Overall, the researchers purpose to find a function that the peaks reflect characteristics of onset.

1.2 Novelty onset detection

In the nearly decade, the technique of deep learning and data-driven approaches make an efficient method to model onset function. As the universal approximation theorem, the neural network can illustrate the ability to map any finite-dimensional discrete space to another arbitrary function [15]. The probability distribution can provide an effective onset function that the property of onset is reflected by the magnitude of probabilistic values while mapping non-onset to a very low value when the system input is spectral of voice. The basic data representation of sound is spectral feature, which can be regarded as an image or a sequence which is an unique identifier of voice, spectrogram, that generate two novel method to onset functions, i.e., onset detection with recurrent neural networks (RNNs) [16] or convolutional neural networks (CNNs) [17]. Moreover, the views on improved sequence representation through long-short term memory neural networks (LSTMs) [18] that further model onset in the signal. Also in MIREX [19] challenge every year, there is a challenge on onset detection, the work of the Böck [20] is considered as the state-of-art methods of universal onset detection [21][22]. However, it is hard to avoid spurious peaks, i.e., false positives, and much more challenging is the detection of onsets in the case of non-percussive music. It is necessary to add a post processing program as Fig 1 to onset detection system after the peak-picking algorithm that could detect the real onsets when it eliminated the false detected onsets.

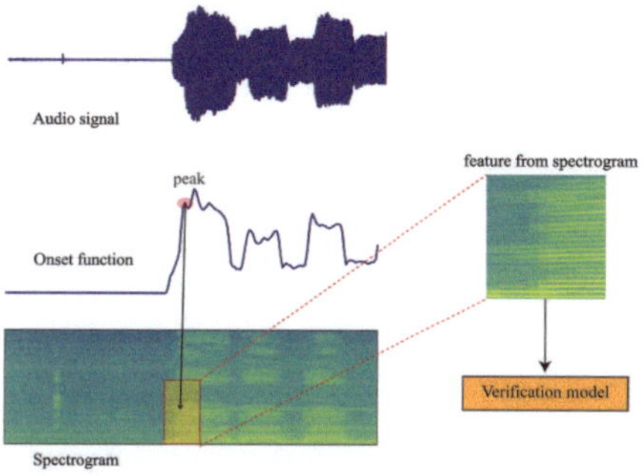

Fig. 1. Onset function and peak-picking algorithm, the verification program aims to reduce false detected onsets according to local feature in spectrogram.

1.3 Our works

The universal onset detection to every kind of music performance is beyond the scope of our research, this paper presents a specific system for solo singing onset detection, and we designed the pipeline with a post-processing step, according to the characters of onset that unstable to the stable frequency distributions in the spectrogram. The purpose of post-processing is to reduce false alarms, that we use a classification model to discern the onset feature in the spectrogram, while the traditional algorithm of onset detection, peaks in the function not always represent onset. It is difficult to describe the difference between onset and false onset in onset function that onsets are always peaks, however, not all peaks are onsets. Onset detection in music signal can deem as a specific scenarios of event detection, which often used to compare with classification. For example, speech recognition is a situation that detection-by-classification, inspired by the research about event detection or classification in music signal [23], we propose a verification step as Fig 1 to reduce false positives of onset detection.

Because of the post-processing, the program can hand over the false peaks picked to verification, which reduce the false detected onsets, we no need to consider complicated onset function and picking algorithm in detection. It solely add post-processing as a verification step that the problem becomes a classification over local feature in the spectrogram. Moreover, the performance of the classification model is directly related to onset results, and the classifier has good performance if the previous procedure output training data in an equal quantity of positive and negative samples. Sequence classifier can be trained and predicted the real onset when we have data and label. Furthermore, the spectrogram is an excellent data representation of audio, and we trained a classifier in the verification step to reduce false alarms.

The detection problem is similar to object detection when it considered a post-processing step, which is a popular topic in the field of computer vision when the spectrogram can be an image of the sound. In the object detection task, a window is used to segment the pixels and then classify the content in the window as a detection output [24]. In our case, the value of onset function can express some characteristics of an onset feature, which is not necessary to detect onsets through the whole spectrogram pixel-by-pixel. The sliding window sampled all the peaks that can be onsets in the spectrogram, when the onset function confirms the possibilities of onset, and then classify them in these places. If it passes the verification, it is a positive sample, i.e., true onset, otherwise, the opposite.

We present an overview on onset detection and our post-processing step in section 1, and explain the deep neural networks as a sequence classification model in post-processing in section 2, the paper presents the spectral-based method of onset detection and modify the peak-picking as our baseline approach, then designs a verification in detection. Section 3, the evaluation dataset to our approach. We finish our discussion in section 4 and 5.

2 Onset detection

The representations of note in the signal are pitch, onset, and offset [25]. Pitch is periodicity description in local scale time, and onset is the start time of a note. Moreover, the transcription system converts the input signal into note sequence in a music score, that intimate the action of musician to listen and record the music. The system can help the singer to compose a work effectively. A standard transcription system aims to determine the note segment in the signal after pitch estimation. The note often begins with a pulse of energy or other physical quantities, yet sometimes the pulse is not indicated obviously in signal data or another form of transform. However, people are sensitive to the onset time when they heard the music, and track beats unconsciously. People can get a perception of beats with onset time and duration of time. That is what singing transcription needs to implement, and finds onsets to determine the note's position in MIDI, i.e., onset detection.

Methods of onset detection can be simplified to two steps in any case of the description of something changes in the signal. Moreover, the essential step is the construction of onset detection to represent the probability of onset in the signal, which can describe the fluctuation of some physical quantity, and derived to the differential function. Most of the research starts here to reduce false detected peaks. A typical and simplified onset detection involves spectral feature of a signal, which is related to the data representation of audio, i.e., spectrogram. We implement the post-processing for the spectral onset function and then use the feature such as spctrogram or Mel Frequency Cepstrum Coefficient (MFCC) in verification.

2.1 Onset function

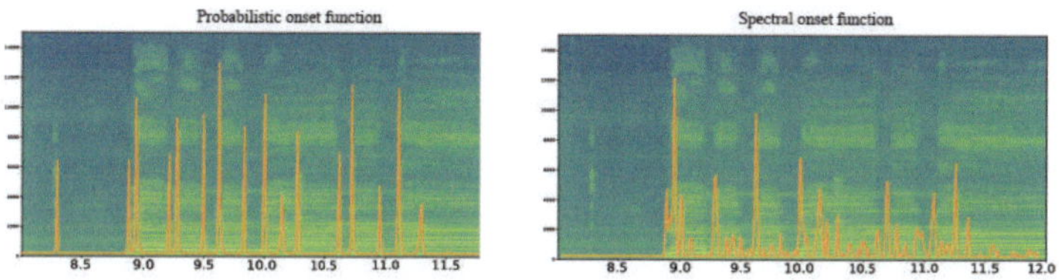

Fig. 2. Onset function with the similar peaks, however noised data occur at the same time, that cause the spurious peaks (double detected peaks) which can't be eliminated by the higher ability of deep neural network (probabilistic onset function).

Onset function describes the probability of onsets in the signal, and the input has time-frequency information when considering the complex-domain model besides energy function. However, the spectrogram leads to those peaks repeatedly occurs in the same places as Fig2. As a consequence, the description of note, the novelty onset function with deep learning model, the value of activation denotes the probability of onset appears, that have higher performance than the traditional model. However, the probabilistic function also generates spurious peaks. This state-of-art method still not present onset perfectly when it converts noised inputs into spurious peaks. The algorithm of onset detection has to pay attention to the no-perception region of audio and eliminate the spurious peaks in onset function.

The detection system extract spectrogram feature from input signal, that better reflects the properties of audio, then using a multivariate function that map vector to one-dimensional data. A type of derivative operator is applied to the multivariate function to get onset function that has higher values at onset time. Subsequently, the peak-picking algorithm is employed to detect the onset candidates for the latter step of verification. However, energy distribution in spectrogram, fluctuations coming from singing may be more visible than energy increase. Moreover, onset detection is concentrated more in the distribution of frequency while voiced sound that related to regular frequency changes are often well detectable in the lower frequency region in the spectrogram, i.e., spectral method. Moreover, the spectral-based method is first to convert the signal into a spectrogram, and then make a function that can capture changes at the onset time. To detect spectral changes in the signal, one computes the difference between subsequent spectral vectors by using a suitable distance measure [10].

To enhance weak spectral components, we apply a logarithmic compression Y to the spectrogram X that $Y = \log(1 + X)$, the horizontal axis of spectrogram is time while vertical axis is frequency, and the depth of colour represent the energy, therefore the spectrogram denote the distribution of energy in frequency at that time. Then onset function by discrete temporal derivative of the spectrum Y is calculated. The onset function transforms a given spectrogram Y into a representation that reflects the different pattern of onset and non-onset. Moreover, the onset functions usually get a local maximum as the onset time, but the opposite is not true that not all local maxima are onsets.

$$F_{onset} = \Delta Spectral = \sum_{k=0}^{K} |Y(n+1,k) - Y(n,k)| \tag{1}$$

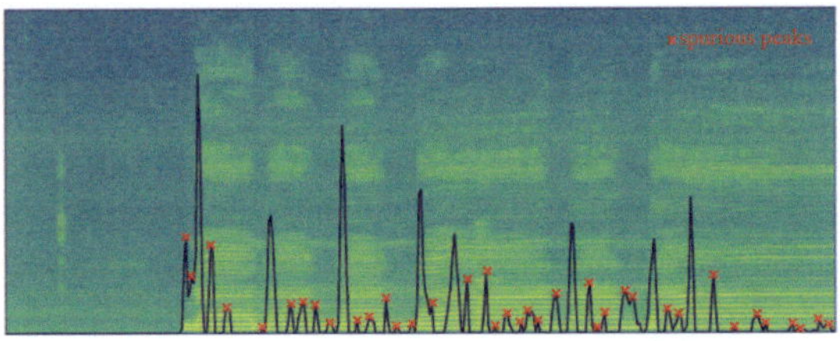

Fig. 3. Spectrogram of a piece of solo singing and onset function. The onsets in spectrogram are equal to peaks in onset function, but not all peaks are onsets,i.e., spurious peaks.

2.2 Peak-picking algorithm

When consider the peak-picking algorithm or the duration between two adjacent onsets, we should not neglect the experience about human hearing. The system has to take account of the relationship between duration of sound and local scale time of peaks. Scientists have researched the relation between threshold and duration for tone pulses, over a wide range of frequencies and duration. Besides, a reasonable range of duration, the ear appears to integrate the energy of the stimulus over time in the detection of short duration tone bursts according to the psychology of hearing [26]. If this were true, the following formula would hold:

$$I \times t = constant \tag{2}$$

where I is the threshold intensity for a tone pulse of duration t. Although there is no evident experiment to prove it [26], this ambiguous formula can give the choices of local time scale as a reference to ensure the parameters of sliding window. The duration t can be proved through the subsequent data analysis results. Peaks with a uncertain duration t in onset function can provide useful onset candidates before the verification. In the absence of the optimal method, we will adjust the parameters of sliding window through experiments to determine a better time interval. A compared group experiment between 100m and 200ms will be set up to find the optimal parameters of duration and overlap.

2.3 Verification

The transcription system matches note patterns in solo singing, and focuses on the front spectrogram of onset in detection component. However, detection is harder than classification, that not solely find the category but the time position in the signal. The verification step takes this problem apart, only focus on positions of onset in the spectrogram, i.e., the start time. Discussions about detection and classification [23] improved detection pipeline with a post-processing step that provides a powerful idea to our scenarios of solo singing. The system finds onsets in local peaks of the function then verification step classifies the adjacent frames of spectrogram according to the time information of peaks. People's hearing is sensitive to onset region, and detection system converts perception peaks as positives, meanwhile the peaks in not perception region became false onsets automatically. For the consideration of feature, the field of speech recognition, a modified feature of technology from GMM-HMM to DNN-HMM [27] indicates that the local MFCC can represent the different phoneme in continuous vector sequence. Moreover, in our research, we trained a classification model as a mechanism to verify the onset feature that local window in spectrogram can be the input of our classification model.

Sequence data and Classification Onset often occurs at a time that unstable to the stable curve of the pitch, when the system computes the F0. The stable pitch in spectrogram can present a clean line which at the beginning time of the note while that unstable pitch not, also the white noise with uneven frequencies distribution. The unbalanced frequency distribution in spectrogram impact the onset function with spurious peaks and results of peak-picking are unreliable. Peaks are onsets, but non-onset in the majority of cases, and the system improve the onset detection pipeline of solo singing with a verification step that converts the detection to a single classification. As Fig 3, each peak with a specified interval to adjacent one, which may be an onset, i.e., it is a candidate to real onset before it classified as positive.

We can get a feature of network input according to the peaks of onset function, localize the window in spectrogram after peak-picking, and classification model in this example is the multi-input, and single-output. Long and short memory neural networks (LSTMs) can model this complex process, and use the cross-entropy loss function.

$$Loss = \sum_{i=1}^{N} y^{(i)} log \bar{y}^{(i)} + (1 - y^{(i)}) log(1 - \bar{y}^{(i)}) \tag{3}$$

In addition, on the open source community, there are already excellent LSTMs framework can be used [28], the basic idea of LSTMs is to use different types of gates to control the flow of information in the network. A complex and sophisticated network unit capable of holding information for a long period of time through a gate structure LSTMs. The argument can decide when to remember input information and when to forget it, and the LSTMs structure can effectively use the context to learn the acoustic characteristics distorted by noise and echo, which can effectively deal with highly unstable noises [27]. Mathematically description of LSTM neurons can be written as follows [29] from the time$T = 1, 2, .., T$:

$$\mathbf{i}_t = \sigma(\mathbf{W}^{(xi)}\mathbf{x}_t + \mathbf{W}^{(hi)}\mathbf{h}_{t-1} + \mathbf{W}^{(ci)}\mathbf{c}_{t-1} + \mathbf{b}^{(i)}) \tag{4}$$

$$\mathbf{f}_t = \sigma(\mathbf{W}^{(xf)}\mathbf{x}_t + \mathbf{W}^{(hf)}\mathbf{h}_{t-1} + \mathbf{W}^{(cf)}\mathbf{c}_{t-1} + \mathbf{b}^{(f)}) \tag{5}$$

$$\mathbf{c}_t = \mathbf{f}_t \cdot \mathbf{c}_{t-1} + \mathbf{i}_t \cdot \mathbf{tanh}(\mathbf{W}^{(xc)}\mathbf{h}_{t-1} + \mathbf{W}^{(hc)}\mathbf{h}_{t-1} + \mathbf{b}^{(c)}) \tag{6}$$

$$\mathbf{i}_t = \sigma(\mathbf{W}^{(xo)}\mathbf{x}_t + \mathbf{W}^{(ho)}\mathbf{h}_{t-1} + \mathbf{W}^{(oi)}\mathbf{c}_t + \mathbf{b}^{(o)}) \tag{7}$$

$$\mathbf{h}_t = \mathbf{o}_t \cdot tanh(\mathbf{c}_t) \tag{8}$$

An onset in the spectrogram are the region before a block of Fig 4, in our approach, the model input is related to the begin component of note. The architecture for sequence-level classification shows that we no longer care about the outputs at each node of neuron but defer the class output to the end of the feature. In other words, a single class label is determined at the very last time-step of the spectrogram of onset, and it is used to backpropagate the class prediction errors. We select 40-dimensions of MFCC that contains information of time-frequency.

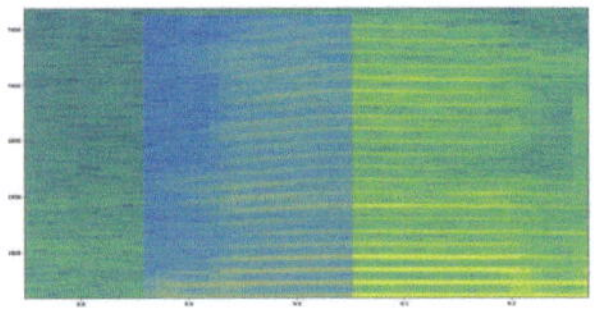

Fig. 4. A note with a onset in front region of the feature.

The multiple inputs and a single output model, sequence-level classification is often leveraged in sentiment analysis. Moreover, in views on the mathematical expression and network structure, the problem we studied is equivalent to sentiment analysis. The system attempts to discover how positive or negative, people are about specific topics by analyzing the content of a sentence, which is the same as the spectral feature of onset or non-onset. For example, one can use a fixed time length of spectrogram (or MFCC when it was filtered) to determine whether or not a an onset. we select the 6 layers of LSTM which has 256 cells in each hidden layer.

3 Evaluation

3.1 Dataset

The testing dataset is from the research SIPTH [7], which consists of 38 melodies sung by men, women, and children. The researchers have been manually annotated the 38 songs. With the sample rate of 44100Hz and a resolution of 16 bits, total duration of the whole dataset is 1154 seconds. And in our approach, course of the action that pick-peaking for feature, the algorithm can generate almost 5000 sample include onsets (positives) and non-onsets (negatives). According to the type of singer, the testing dataset divides into three classifications: 14 melodies of traditional children songs, 13 songs of adult male and 11 melodies of adult female.

The corresponding dataset, training set is from the public dataset, MTG-QBH [30] which contain some elements from the testing dataset, we picked the rest 94 songs that were not in the test dataset (i.e., 11 melodies of adult male and 13 melodies of adult female), the total duration of the whole dataset is 42 minutes and 46 seconds (about 2566 seconds), which can generate more than 10000 features and labels for onset or non-onset. We annotated onset time in every piece of songs when we heard and saw some event changes in spectrogram, which do not focus on the details of the phonemes in speech.

Onset features and labels in training dataset can be created by a simple way that we label the frequency distribution changed suddenly as Fig 4, the onset is a description of a starting time of note, and the subsequence vector in spectrogram become more stable than the left-location. We can only hear pitches where the sound is stable, we select the local feature when the peak-picking algorithm output the almost peaks and compare the onset time with our ground-truth, record it as a positive if the absolutely difference-value is less than 50ms. The ground-truth of listener in visualization is more accuracy than hearing and recording. Features are located by peaks, and the labels occur, which makes dataset simple.

3.2 Classification and Onset Detection

The results of detected onset times will be compared with the ground-truth files from the SIPTH [7]. For a given ground-truth onset time, if there is a detection in a tolerance time-window (+/-50 ms) around it, then it is considered as a correct detection [19], or opposite when the time-window is greater than tolerance. Moreover, the effective evaluation parameters are precision, recall and F-measure. In our example, a tolerance time-window of detection and labels were introduced to evaluate positive, if the detection time less than the tolerance time, then it is a positive, if not, negative.

4 Result

4.1 Peak-picking in sliding window

Peak-picking algorithm detects onsets by a threshold. In our plan, we adopt local maximum of function by a sliding window, and not focus too much on represent onset, because there are subsequent verification steps. According to our previous analysis, in fact, there is an imprecise approximate time range between each onset. Although we don't know what this range of duration is, this principle can provide us with some parameters between two successive onsets. Based on the previous analysis, the window length is smaller, then recall rate is higher with lower precision.

Suppose we set up equal intervals from 100 ms to 200 ms with overlap 50% of intervals, and find the optimal parameters to peaks, which can determine the quality of data for training networks. The training dataset should contain onsets as much as possible, that is, recall should be as greater as possible, indicating that the maximum number of samples can be detected. Meanwhile, the training dataset should make positive and negative samples as balanced as possible. In the case that training data and labels contain noises, we set up trials to find as many onsets as possible. It's probably the best time window between 100m and 200ms, overlap is 50% length of the window. The results are as follows:

Table 1. Evaluation on Peak-picking

local window(ms)	100	120	140	160	180	200
precision	0.305	0.360	0.414	0.462	0.511	0.546
recall	0.951	0.939	0.920	0.915	0.904	0.895
F1-measure	0.462	0.520	0.572	0.614	0.653	0.678

4.2 Baseline and Verification

We use onset function and peak-picking algorithm as baseline, and the experiments show that the verification steps can reduce false alarm, thus improving the final detection rate. Choose the 140ms window length with 50% overlap, then train the classifier in the following steps, to solve the false detection, and then find the real onset. When training network, the amount of dataset is not very large, it is easy to produce over-fitting, so we adopt the operation of early termination, that is, when the accuracy of training network reaches more than 90%, the training process is stopped manually. Finally, the network can only correctly detect more than 80% onsets that peak-picking about 92% onsets.

4.3 Result and Discussion

We made some modifications in spectral function, the local maximum is selected to determine peaks in our baseline approach, but it doesn't change the principle that eliminated a part of spurious peaks, the result is not effective and there are still many false positives, that is, the precision are lower than some methods in other research [17, 16]. However, we promote the peak-picking algorithm, rather than construct complicated function when the values are greater than classical threshold method which impact the performance that two consecutive onset are likely to appear, the local maximum within a certain range will not. This range of local time can determine the shortest duration of a note by analysing some common singing methods and the time value of the note.

In practice, the detection rate affected by the performance of classification model as we add a verification as a post-processing step. The classification model with multiple inputs and single outputs that achieves 0.90 to 0.94 accuracy on the training set and more than 0.9 on the test set. When checking the results, it can be found that the classifier can only find about 90% of onsets, which are extracted from the original data by onset function and peak-picking. Before verification, the precision is only 0.41 then the value can be improved to 0.92. In the verification step, the classifier achieves 90% correct classification of samples. After verification, precision is 0.92, and recall is 0.85. As we expected, about 90% onset can be found in the previous steps of classical methods, but there are many negative samples in the results that impact the precision. The verification step reduced false alarms in our baseline approach, and we can estimate the result roughly, as we expected the peak-picking find the 92% onsets, then classifier classifies 90% of the samples correctly, i.e., eliminate spurious peaks, that the final recall reaches about 0.85.

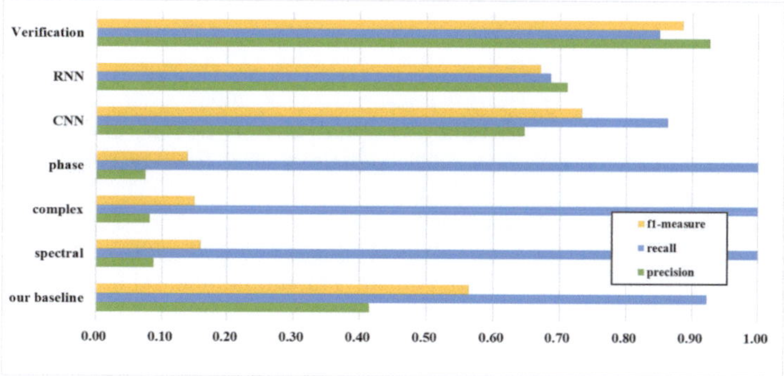

Fig. 5. Onset detection evaluation, the CNN or RNN method is the probabilistic onset function which is the state-of-art method of universal onset detection while the phase, complex and spectral are the methods in common textbooks of music information retrieve (MIR)

5 Conclusion

The transcription system, which has a generally developed framework nowadays, consists of pitch tracking and note segmentation. However, the system details are different in the face of multiple forms of music performance. The result of note recognition is related to note segmentation, and the onset detection in note segmentation is the core task of the whole problem. We have implemented the typical method of onset detection in the textbook [10], which could be seemed as a baseline approach before verification. And we also adopted the combination method of onset detection and verification that is optimised for solo singing by using the spectrogram that is the basis of speech recognition.

Our approach of singing onset detection is to design a specific system, and then improve problem by training the similar dataset, the solo singing transcription system can use the detected sequence of the time position of onset in the music signal, that reproduce singing voice to note sequence. In the verification step, the performance of classification model is directly associated to the final onsets, and higher performance of detection can affect the consequence of note recognition in singing that will be a very important fundamental research in music transcription.

Music transcription is very sensitive to onset time, previous studies of transcription system, onset detection is widely used but not focus on note segmentation that could not have great performance, and the pitch estimation can output reliable pitch sequence while that previous system not detection onset in a correct position of music signal, which lead to poor transcription of notes. The research SPITH [7] mentions that onset detection can be used to improve the note recognition, which is the most relevant part of the whole transcription system, and the system should be optimised for specific situations.

References

1. Matti Ryynänen. *Singing Transcription.* Springer, 2006.
2. J. P. Bello, L. Daudet, S. Abdallah, C. Duxbury, M. Davies, and M. B. Sandler. A tutorial on onset detection in music signals. *IEEE Transactions on Speech and Audio Processing*, 13(5):1035–1047, 2005.
3. Yongwei Zhu, Mohan S Kankanhalli, and Changsheng Xu. Pitch tracking and melody slope matching for song retrieval. pages 530–537, 2001.
4. De Cheveigné Alain and Kawahara Hideki. Yin, a fundamental frequency estimator for speech and music. *Journal of the Acoustical Society of America*, 111(4):1917–30, 2002.
5. Matthias Mauch and Simon Dixon. Pyin: A fundamental frequency estimator using probabilistic threshold distributions. In *IEEE International Conference on Acoustics*, 2014.
6. Camacho Arturo and John G Harris. A sawtooth waveform inspired pitch estimator for speech and music. *Journal of the Acoustical Society of America*, 124(3):1638, 2008.
7. Emilio Molina, Lorenzo J. Tardon, Ana M. Barbancho, and Isabel Barbancho. Sipth: Singing transcription based on hysteresis defined on the pitch-time curve.

IEEE/ACM Transactions on Audio Speech and Language Processing, 23(2):252–263, 2015.
8. Hoon Heo and Kyogu Lee. Robust singing transcription system using local homogeneity in the harmonic structure. *Ieice Transactions on Information and Systems*, 100(5):1114–1123, 2017.
9. Rodger J. Mcnab, Lloyd A. Smith, and Ian H. Witten. Signal processing for melody transcription. *Proc.australasian Computer Science Conf.–*, pages 301–307, 1995.
10. Meinard Müller. *Fundamentals of Music Processing*. Springer, 2015.
11. A. Holzapfel, Y. Stylianou, A. C. Gedik, and B. Bozkurt. Three dimensions of pitched instrument onset detection. *IEEE Transactions on Audio, Speech, and Language Processing*, 18(6):1517–1527, Aug 2010.
12. Ruohua Zhou, M. Mattavelli, and G. Zoia. Music onset detection based on resonator time frequency image. *Trans. Audio, Speech and Lang. Proc.*, 16(8):1685–1695, nov 2008.
13. Paul Masri. *Computer modelling of sound for transformation and synthesis of musical signals*. PhD thesis, University of Bristol, 1996.
14. Sebastian Böck and Gerhard Widmer. Local group delay based vibrato and tremolo suppression for onset detection. In *ISMIR*, pages 361–366. Citeseer, 2013.
15. Yann LeCun, Yoshua Bengio, and Geoffrey Hinton. Deep learning. *Nature*, 521:436–444, 2015.
16. Sebastian Böck, Andreas Arzt, Florian Krebs, and Markus Schedl. Online real-time onset detection with recurrent neural networks. In *Proceedings of the 15th International Conference on Digital Audio Effects (DAFx-12), York, UK*, 2012.
17. Jan Schlüter and Sebastian Böck. Musical onset detection with convolutional neural networks. In *6th international workshop on machine learning and music (MML), Prague, Czech Republic*, 2013.
18. Florian Eyben, Sebastian Böck, Björn Schuller, and Alex Graves. Universal onset detection with bidirectional long-short term memory neural networks. In *Proc. 11th Intern. Soc. for Music Information Retrieval Conference, ISMIR, Utrecht, The Netherlands*, pages 589–594, 2010.
19. MIREX. Audio onset detection. https://www.music-ir.org/mirex/wiki/, 2019.
20. MIREX. Audio onset detection. https://nema.lis.illinois.edu/nema_out/mirex2018/results/aod/summary.html, 2018.
21. Sebastian Böck, Jan Schlüter, and Gerhard Widmer. Enhanced peak picking for onset detection with recurrent neural networks. In *International Workshop on Machine Learning and Music*, 2013.
22. Sebastian Böck, Florian Eyben, Björn Schuller, and Technische Universität München. Mirex 2010 submission: Onset detection with bidirectional long short-term memory neural networks. *Proc ISMIR*, 2013.
23. Huy Phan, Philipp Koch, Fabrice Katzberg, Marco Maass, Radoslaw Mazur, Ian McLoughlin, and Alfred Mertins. What makes audio event detection harder than classification? In *2017 25th European Signal Processing Conference (EUSIPCO)*, pages 2739–2743. IEEE, 2017.
24. Junge Zhang, Zhao Xin, Yongzhen Huang, and Kaiqi Huang. Semantic windows mining in sliding window based object detection. In *International Conference on Pattern Recognition*, 2012.
25. A. Klapuri and M. Davy. Signal processing methods for music transcription. 2006.
26. Brian C. J Moore. An introduction to the psychology of hearing, 5th ed. *Archives of Otolaryngology*, 103(12):745–746, 1977.
27. Deng Li and Dong Yu. *Automatic Speech Recognition: A Deep Learning Approach*. Springer, 2014.

28. François Chollet et al. Keras. 2015.
29. F. A. Gers, J Schmidhuber, and F Cummins. Learning to forget: continual prediction with lstm. *Neural Computation*, 12(10):2451–2471, 2000.
30. Serrá Joan Salamon, Justin and Emilia Gómez. Tonal representations for music retrieval: from version identification to query-by-humming. *International Journal of Multimedia Information Retrieval*, 2(1):45–58, Mar 2013.

Transfer learning for music classification and regression tasks using artist tags

Lei Wang[1], Hongning Zhu[1], Xulong Zhang[1], Shengchen Li[2], and Wei Li[13]

[1] School of Computer Science and Technology, Fudan University, Shanghai, China
[2] Institute of information photonics and optical communications, Beijing University of Posts and Telecommunications, Beijing, China
[3] Shanghai Key Laboratory of Intelligent Information Processing, Fudan University, China

Abstract. In this paper, a transfer learning method that exploits artist tags for general-purpose music feature vector extraction is presented. The feature vector extracted from the last convolutional layer in a deep convolutional neural network (DCNN) trained with artist tags is showed for music classification and regression tasks. Not only are artist tags adequate in the music community, therefore easy to be gathered, but also contain much high-level abstract information about the artists and the music audio released by the artists. To train the network, a dataset containing 33903 30-second clips, annotated with artist tags was created. The model is trained to predict the artist tags from audio content first in the proposed system. Then the model is transferred to extract the features that are used to perform music genre classification and music emotion recognition tasks. The experiment results show that the features learned using artist tags under the context of transfer learning are able to be effectively applied in music genre classification and music emotion recognition tasks.

Keywords: Transfer learning, Music emotion recognition, Music genre classification

1 Introduction

Recently deep learning have invaded music information retrieval (MIR) community, getting remarkable performance in many music-related tasks such as automatic music transcription [1–3], music recommendation [4], music emotion recognition [5–7] etc. The working principle of deep learning is that deep learning models are able to extract patterns out of a lot of training data. However, annotating music demands professionally trained musicians, thus it is hard to establish a massive database of music labels. Looking to music emotion recognition, the fact is that the process of labeling the emotion labels for music excerpts by music experts is expensive and time-consuming. Lack of adequate training data hinders the application of deep learning in MIR research to a certain extent.

Transfer learning means that the model trained on a dataset is then transferred to target tasks, leading the knowledge learned in large datasets to be

H. Li et al. (eds.), *Proceedings of the 7th Conference on Sound and Music Technology (CSMT)*, Lecture Notes in Electrical Engineering 635,
https://doi.org/10.1007/978-981-15-2756-2_7

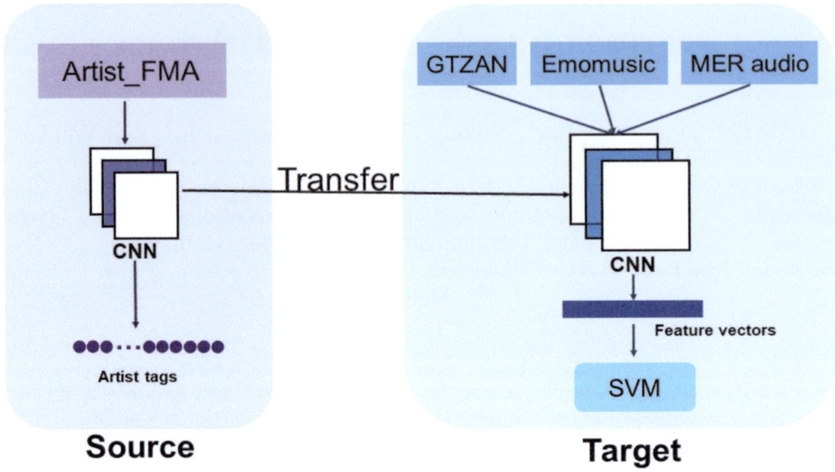

Fig. 1. A diagram of framework, The deep CNN model trained with artist tags is described in the left part of diagram. In the right part of diagram, the pre-trained model that used as a feature extractor is then transferred to target datasets. The extracted feature vectors would be used to train a support vector machine.

transferred into small datasets. The model is often trained with adequate data in the source task and transferred to the target task, which makes the trainable parameters in the target dataset significantly reduced. There are many transfer learning examples in computer vision and natural language processing. The network trained on the ImageNet dataset [8] for image classification leads to significantly improved results for object and action classification [9]. Pre-trained word embedding which is trained on a large dataset is another example of transfer learning [10].

Artist tags contains abstract information from the perspective of human describing the styles of an artist such as music emotion (sad, happy, quiet), instrumentations (guitar, piano, vocal) and music genres (jazz, rock, country, blues) that the artist is skilled at. On the one hand, these tags are labeled by listeners when they listen to songs that released and covered by artists. Therefore, tags contain rich high-level information about artists from the perspectives of the listeners. Furthermore, tags can be regarded as terms that describe the diverse music styles of artists. On the other hand, these tags are easily gathered since they are sufficient available resources in music community. In this paper, automatic artist tagging is defined as a multi-label prediction task that intends to assign the aforementioned types of tags for the artist using the audio signal, which is split from the songs released/covered by the artist. In the proposed transfer learning method, automatic artist tagging is selected as the source task and a DCNN model is built for automatic artist tagging. Furthermore, the DCNN model with

Layer	Output shape
Mel-spectrogram	(128, 129)
128 kernel, Conv1d	
4 Maxpooling	(128, 32)
128 kernel, Conv1d	
4 Maxpooling	(128, 8)
128 kernel, Conv1d	
4 Maxpooling	(128, 4)
128 kernel, Conv1d	
2 Maxpooling	(128, 2)
256 kernel, Conv1d	
Global average pooling	(256)
Fully-connected layer	(50)

Table 1. The structure of the network

the trained parameters is transferred to target datasets, which is used to extract feature vectors. Since the artist tag is a song-level descriptor, it is proper to transfer the pre-trained model to the song-level classification/regression tasks. Therefore, music genre classification and music emotion recognition are chosen as the target tasks.The underlying motivation is that if a model is trained to tag artists well and discriminate different styles for artists from music, the model can be assumed to learn different characteristics of music based on various aspects of artist tags such as emotion, instrumentations, and music genres, etc.

For the rest of this paper, the transfer learning method are drawn in the section 2. How to make the proposed dataset is introduced in Section 3. The experiment details and results are described in the following section. The conclusions and future work are drawn in the last section.

2 Proposed method

In this section, the transfer learning framework (Fig.1) is described. A DCNN model is built to be trained with artist tags. The network with trained weights is transferred to target datasets and is used as a feature extractor. The feature vectors extracted from the last convolutional layer are served as the input to support vector machine (SVM). This proposed method is evaluated on three target datasets: MER audio, Emomusic, and Gtzan.

2.1 Preprocess of audio

The input to the network consists of mel-spectrograms with 128 frequency bins, which is a kind of representation of audio based on time and frequency dimension. The mel-spectrogram is chosen because it is psychologically similar to the way of perceiving sound by human. To get the mel-spectrograms, FFT window size (n-fft) and hop length are set as 1024 and 512 in order, and then the frequency axis which is obtained by performing short-time Fourier transform is converted to 128 Mel-frequency bins. Finally, we perform a logarithmic transform on the magnitude of mel-spectrograms. The audio transform to mel-spectrograms is implemented by using `Librosa`[4] [11].

2.2 Model for artist tagging

A neural network (Table 1) contained five 1D-Convolutional layers and one fully connected layer is built. In the proposed neural network, zero padding with *2* is added to both sides of the input in the first four convolutional layers. Rectified Linear Units (ReLU) is set as the activation function in all hidden layers, and Sigmoid is used as the activation function after fully connected layer. Batch Normalization [12] is used after each convolutional layer, Dropout [13] is attached at the end of the last convolutional layer to mitigate overfitting. The size of 1D-MaxPooling layers is 4, 4, 4, 2 in order and a global average pooling layer is used after the last convolutional layer.

2.3 Feature transfer and extraction

After training on the source dataset, the pre-trained network is transferred into the target datasets used as a feature extractor. The input of the network is a mel-spectrogram of 3-second fragments of audio. In the section of feature extraction, to get a representation for a longer clip on the target datasets, the spectrogram of clips whose duration is short than 30 seconds is expanded to 30 seconds by duplicating the original music audio. Then the 30-second clip is split to 3-second fragments. Finally, the representations of 10 fragments gathered from the activation of the last layer are averaged into a feature vector as the representation of a 30-second clip. The representations of clips are used for different classification and regression tasks.

2.4 Classification

Support vector machine (SVM) has been the most popular classifier in MIR works [14, 15], so we select it as the classifier removing classifier dependency in this paper. SVM with parameters obtained by doing grid-search on the validation set are selected as the standard classifier. Parameters of the bandwidth and the penalty are searched in the same range as literature [15]. The implement of SVM and 10-fold cross-validation are based on `Scikit-learn`[5] [16].

3 Datasets

In this section, how to make the proposed source dataset and the details of the source dataset used to train the neural network are introduced. Three target datasets which are divided into two categories: (music emotion recognition, and music genre classification) are used to evaluate the proposed method.

[4] https://github.com/librosa/librosa
[5] https://scikit-learn.org/stable/index.html

3.1 Source dataset

Data collect For training the neural network, the source dataset which we name artist_FMA whose all songs files are crawled from *freemusicarchive.org*[6] (FMA) [17] under Creative Commons licenses is proposed. The artists' names of songs in the dataset were collected at the same time. In the process of collection, artists whose songs are more than 20 were crawled from FMA. After collection, the songs whose duration is less than 60 seconds are removed. And then they are split to 30-second clips. The dataset consists of 498 artists and 47870 clips.

`Last.fm`[7] is a music community and the active users in the community usually add tags to their favorite artists or songs when they listen to these artists and songs. There are two reasons to choose automatic artist tagging which is similar to automatic music tagging task [18] as the source task: *i*) there are enough available artist tags. *ii*) all tags associated with each artist usually reflect all kinds of characteristics of the artist. And the top tags can reflect the most characteristics of the artists. `Pylast`[8] is utilized to call `Last.fm` API with the artist's name, and all associated tags for artists are collected from the community. However, there is a considerable number of artists collected from FMA that have ordinary reputation compared to famous artists like Lady Gaga and Taylor Swift and whose songs are unpopular with music listeners. We can not gather enough tags for the corresponding artists in this case. As a result, the final dataset (artist_FMA) used to train the neural network consists of 352 artists, 33903 clips, and the associated tags. Fifty artist tags [18] which cover music genres (rock, Hip-Hop, electronic, experimental, etc.), eras (from 60s to 00s), moods labels, and instrumentations (piano, guitar) are selected to filter all collected tags, and the 50-dimension vectors are obtained as the label of the artists.

Dataset split For split strategy of artist_FMA dataset in this paper, three single train/validation/test split are designed, obeying the following strategies. The train/validation/test sets make up roughly 70/20/10 percent of the dataset (in the amount of clips)

1) Artist-level_split: No clips split from the same artist should appear in more than one split set. For example, if all clips split from Tailor Swift's songs, they would appear in the training set, rather than appearing in validation and test set.
2) Song-level_split: No clips from the same song should appear in more than one split set. For example, if five clips split from *YouNeedtoCalmDown*, and these five clips would be used to train the model, rather than being used to validate or test the model.
3) Shuffle_split: Given the model is only used as a feature extractor and we will not report the automatic artist tagging results to compare with other methods, train/validat-ion/test set should be made up by random split. It is

[6] https://freemusicarchive.org/
[7] https://www.last.fm
[8] https://github.com/pylast/pylast

possible that clips from the same song would appear in more than one split set.

3.2 Target datasets

Three target datasets which can be categorized as music emotion recognition (dataset 1 and 2), music genre classification (dataset 3) are chosen to evaluate the features learned using artist tags .

1) Emomusic [19]: The 744 excepts were continuously annotated on the arousal and valence plane with Russell's model [20]. The representation vectors extracted from the model are evaluated by predicting the value of arousal and valence separately in this dataset. Note that only the first 30 seconds per excepts is used. There is similar processing in the literature [15].
2) MER audio [21]: The music data (30-second audio clips) and metadata (the mood tags) were obtained from AllMusic API and the mood label values are obtained by Warriner's list [22] which containing 13915 English words with AV ratings according to Russell's model [20]. The dataset contains 900 clips, annotated into 4 quadrants.
3) Gtzan [23]: The Gtzan dataset is a benchmark dataset that contains 10 genres for music genre classification.

4 Experiment

4.1 Artist tagging

Automatic artist tagging is a multi-label classification task, since an artist can be tagged with multiple tags. There is tremendous difference between automatic tagging with other audio classification problems such as music emotion recognition and music genre classification that are usually considered as one label prediction tasks. Multi-label prediction is more complicated than one label prediction, because the search space of the multi-label prediction is an exponentially increase contrast to one label prediction. For example, considering there are N labels to predict, a classifier that predicts one label only predicts a case among N kinds of cases while a classifier that predicts multi-label should predict a case from 2^N cases in aggregation.

However, the tag vectors tend to be sparse (too much false (0) in tag vectors) for most artists, which makes accuracy or mean square error inappropriate as a measurement. Thus the Area Under an ROC (Receiver Operating Characteristic) Curve abbreviated as AUC is often selected as a measurement for automatic tagging tasks.

With three different split strategies, the model is trained for 200 epochs by using the artist_FMA dataset. Binary-Cross-Entropy is set as loss function and Adam is chosen to optimize the loss. The model that gets the best AUC score in the validation set is used for evaluation.

Table 2. The training results of the model by using artist tags and the summary of performances of feature vector extracted from the pre-trained model. Accuracy is used as measurement in the MER audio and Gtzan dataset, and the coefficient of determination (r^2) is used as the metric for arousal/valence prediction in Emomusic. The best result is marked in bold.

Split strategy	AUC	MER audio	Emomusic arousal	Emomusic valence	GTZAN
Artist-level split	0.8526	0.5360	0.4792	0.0981	0.4840
Song-level split	0.8978	**0.6293**	0.5170	0.2667	0.6410
shuffle split	0.9248	0.6288	**0.5345**	**0.3291**	**0.6650**

4.2 Experiment results and Discussion

The experiment results are summarized in Table 2. Shuffle split and Song-level split outperform artist-level a large margin in three target datasets. Actually, it is difficult to train the neural network in an artist-level split. In artist-level split, the AUC score was starting decreasing or oscillating in early epochs while the loss decreased. The experiment result shows that artist-level split is an inappropriate split strategy. We consider two possible reasons, *i*) there are few commonalities between artists in the artist_FMA dataset, As a result, the neural network did not learn much about these commonalities using one part of artist tags to predict another part of artist tags. *ii*) 352 artists are still not enough for artist-level split strategy, and more artists and songs are needed to train the network.

The scores achieved in target datasets prove that song-level split and shuffle split are effective split strategies, And the model finally achieves 89.71% and 92.48% AUC score respectively. The Shuffle split achieves the best performances in Emomusic and Gtzan dataset and comparable result in MER audio compared with the Song-level split. Although the motivation of the Shuffle split is stated in section 4.1.2, we hypothesis that the performance of the pre-trained model may be lower than song-level split. However, the hypothesis is rebutted by the results. We infer that 30 clips from the same song are different due to music has a period structure. Fig.2 shows the performances of music genre classification with three split methods in Gtzan dataset. Even though the neural network was trained to tag for artist, the representations picked up by the model can be used to discriminate music genres well. *Classical* label does not exist in top 50 tags, however, the feature vectors extracted from the transferred model can differentiate the classical music from other genres of music.

5 Conclusions and Future work

A dataset artist_FMA is proposed that contains artist tags and audio, where a DCNN is trained as a feature extractor. The knowledge learned by the pre-trained model is transferred to three small target datasets. The experiment results show that the knowledge learned from artist tags could be effectively applied in various music-related tasks. Transfer learning has a promising future

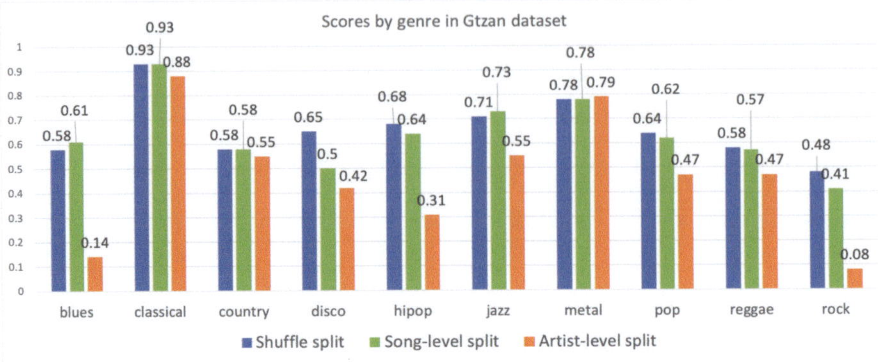

Fig. 2. Summary of performances of three different split strategies.

in a lot of MIR problems since it could be an approach to deal with the situations where there are limited training data to use.

As future work, we firstly plan to improve the artist_FMA dataset by increase the number of artists incorporated in dataset. More training data are able to improve the pre-trained model. Secondly, we consider giving different weights to different tags in the label vectors according to the number of tags. The label vectors with weights setting may improve the pre-trained model.

6 Acknowlegement

This research was supported by NSFC 61671156.

References

1. Hawthorne C, Elsen E, Song J, Roberts A, Simon I, Raffel C, ... Eck D (2017) Onsets and frames: Dual-objective piano transcription. arXiv preprint arXiv:1710.11153
2. Hawthorne C, Stasyuk A, Roberts A, Simon I, Huang CZA, Dieleman S, ... Eck D (2018) Enabling factorized piano music modeling and generation with the MAESTRO dataset. arXiv preprint arXiv:1810.12247
3. Cheng T, Mauch M, Benetos E, Dixon S (2016) An attack/decay model for piano transcription. In: Conference of the International Society for Music Information Retrieval, pp 584-590, New York, America
4. Van den Oord A, Dieleman S, Schrauwen B (2013) Deep content-based music recommendation. In: Advances in neural information processing systems, pp 2643-2651, Lake Tahoe, America
5. Delbouys R, Hennequin R, Piccoli F, Royo-Letelier J, Moussallam M (2018) Music mood detection based on audio and lyrics with deep neural net. arXiv preprint arXiv:1809.07276
6. Weninger F, Eyben F, Schuller B (2014) On-line continuous-time music mood regression with deep recurrent neural networks. In: 2014 IEEE International Conference on Acoustics, Speech and Signal Processing (ICASSP), pp 5412-5416, IEEE, Florence, Italy

7. Li X, Tian J, Xu M, Ning Y, Cai L (2016) DBLSTM-based multi-scale fusion for dynamic emotion prediction in music. In: 2016 IEEE International Conference on Multimedia and Expo (ICME), pp 1-6, IEEE, Yogyakarta, Indonesia
8. Deng J, Dong W, Socher R, Li LJ, Li K, Fei-Fei L (2009) Imagenet: A large-scale hierarchical image database. In: 2009 IEEE conference on computer vision and pattern recognition, pp 248-255, Anchorage, America
9. Oquab, M, Bottou, L, Laptev, I, Sivic J (2014) Learning and transferring mid-level image representations using convolutional neural networks. In: Proceedings of the IEEE conference on computer vision and pattern recognition, pp 1717-1724, Columbus, America
10. Mikolov T, Chen K, Corrado G, Dean J (2013) Efficient estimation of word representations in vector space. arXiv preprint arXiv:1301.3781
11. McFee B, Raffel C, Liang D, Ellis DP, McVicar M, Battenberg E, Nieto O (2015). librosa: Audio and music signal analysis in python. In: Proceedings of the 14th python in science conference (Vol. 8)
12. Ioffe S, Szegedy C (2015) Batch normalization: Accelerating deep network training by reducing internal covariate shift. arXiv preprint arXiv:1502.03167
13. Hinton GE, Srivastava N, Krizhevsky A, Sutskever I, Salakhutdinov RR (2012) Improving neural networks by preventing co-adaptation of feature detectors. arXiv preprint arXiv:1207.0580
14. Park J, Lee J, Park J, Ha JW, Nam J (2018) Representation learning of music using artist labels. In: Proceedings of the 19th International Conference on Music Information Retrieval, pp 717-724, Paris, France
15. Choi K, Fazekas G, Sandler M, Cho K (2017) Transfer learning for music classification and regression tasks. arXiv preprint arXiv:1703.09179
16. Pedregosa F, Varoquaux G, Gramfort A, Michel V, Thirion B, Grisel O, ... Vanderplas J (2011). Scikit-learn: Machine learning in Python. InL Journal of machine learning research, 12(Oct):2825-2830
17. Defferrard M, Benzi K, Vandergheynst P, Bresson X (2016) Fma: A dataset for music analysis. arXiv preprint arXiv:1612.01840
18. Choi K, Fazekas G, Sandler M (2016) Automatic tagging using deep convolutional neural networks. arXiv preprint arXiv:1606.00298
19. Soleymani M, Caro MN, Schmidt EM, Sha CY, Yang YH. (2013) 1000 songs for emotional analysis of music. In: Proceedings of the 2nd ACM international workshop on Crowdsourcing for multimedia, pp 1-6. ACM, Barcelona, Spain
20. Russell JA (1980) A circumplex model of affect. In: Journal of personality and social psychology, 39(6):1161.
21. Panda R, Malheiro R, Paiva RP (2018), In: Musical Texture and Expressivity Features for Music Emotion Recognition, pp 383-391, Paris, France
22. Warriner AB, Kuperman V, Brysbaert M (2013) Norms of valence, arousal, and dominance for 13,915 English lemmas. Behavior research methods, 45(4):1191-1207
23. Tzanetakis G, Cook P (2002) Musical genre classification of audio signals. In: IEEE Transactions on speech and audio processing, 10(5):293-302

Part III
Computational Musicology

MG-VAE: Deep Chinese Folk Songs Generation with Specific Regional Styles

Jing Luo[1], Xinyu Yang[1], Shulei Ji[1], and Juan Li[2]

[1] School of Computer Science and Technology, Xi'an Jiaotong University, Xi'an 710049

[2] Center of Music Education, Xi'an Jiaotong University, Xi'an 710049

Abstract. Regional style in Chinese folk songs is a rich treasure that can be used for ethnic music creation and folk culture research. In this paper, we propose MG-VAE, a music generative model based on VAE (Variational Auto-Encoder) that is capable of capturing specific music style and generating novel tunes for Chinese folk songs (*Min Ge*) in a manipulatable way. Specifically, we disentangle the latent space of VAE into four parts in an adversarial training way to control the information of pitch and rhythm sequence, as well as of music style and content. In detail, two classifiers are used to separate style and content latent space, and temporal supervision is utilized to disentangle the pitch and rhythm sequence. The experimental results show that the disentanglement is successful and our model is able to create novel folk songs with controllable regional styles. To our best knowledge, this is the first study on applying deep generative model and adversarial training for Chinese music generation.

Keywords: Music Generation,Disentangled Latent Representation, Chinese Folk Songs, Regional Style

1 Introduction

Creating realistic music pieces automatically has always been regarded as one of frontier subjects in the field of computational creativity. With recent advances in deep learning, deep generative model and its variants have been widely used in automatic music generation [1][2]. However, most of deep composition methods focus on Western music rather than Chinese music. How to employ deep learning to model the structure and style of Chinese music is a challenging but novel problem.

Chinese folk songs, an important part of traditional Chinese music, are improvised by local people and passed on from one generation to the next orally. Folk tunes from the same region exhibit similar style while tunes from different areas present different regional styles [3][4]. For example, the songs named *Mo Li Hua* have different versions in many areas of China and show various mu-

H. Li et al. (eds.), *Proceedings of the 7th Conference on Sound and Music Technology (CSMT)*, Lecture Notes in Electrical Engineering 635,
https://doi.org/10.1007/978-981-15-2756-2_8

sic styles, though they share the same name and similar lyrics [3] . The regional characteristics of Chinese folk songs are not well explored and should be utilized to guide automatic composition for Chinese folk tunes. Furthermore, folk song composition based on regional style provides abundant potential materials for Chinese national music creation, and promotes the spread and development of Chinese national music and even Chinese culture in the world.

There are lots of studies on music style composition of Western Music[5]. However, few studies employ deep generative model for Chinese music composition. There is a clear difference between Chinese and Western music. Unlike Western Music, which focuses on the vertical structure of music, Chinese music focuses on the horizontal structure, i.e., the development of melody, and the regional style of Chinese folk songs is mainly reflected in its rhythm and pitch interval patterns [6].

In this paper, we propose a deep music generation model named MG-VAE to capture regional style of Chinese folk songs (*Min Ge*) and create novel tunes with controlled regional style. Firstly, a MIDI dataset with more than 2000 Chinese folk songs covering six regions is collected. After that, we encode the input music representations to the latent space and decode the latent space to reconstruct music notes. In detail, the latent space is divided into two parts to present the pitch features and rhythm features, namely, *pitch variable* and *rhythm variable*. Then we further divide the pitch latent space into *style variable* part and *content variable* part to present style feature and style-less feature in pitch variable, the same operation is launched in rhythm variable. In order to capture the regional style of Chinese folk songs precisely and generate regional style songs in controllable way, we propose a method based on adversarial training for disentanglement of the four latent variables, where temporal supervision is employed in the separation of pitch and rhythm variable, and label supervision is used for the disentanglement the style and content variable. The experimental results and visualization of latent spaces show that our model is effective to disentangle latent variables and is able to generate folk songs with specific regional style.

The rest of the paper is structured as follows: after introducing related work on deep music generation in Section 2, we present our music representations and model in Section 3. Section 4 describes the experimental results and analysis of our methods. Conclusions and future work are presented in Section 5.

2 Related Work

RNN (Recurrent Neural Network) is one of the most earliest models introduced into the domain of deep music generation. Researchers employ RNNs to model the music structure and generate different formats of music, including monophonic folk melodies [7], rhythm composition [8], expressive music performance [9], multi-part music harmonization [10]. Other recent studies have started to

[3] The Chorus of *Mo Li Hua Diao* from various regions in China by Central National Orchestra: http://ncpa-classic.cntv.cn/2017/05/11/VIDEEMEg82W5MuXUMM1jpEuL170511.shtml.

combine convolutional structure and explore using VAE, GAN (Generative Adversarial Network) and Transformer for music generation. MidiNet [11] and MuseGAN [12] combine CNN (Convolutional Neural Network) and GAN architecture to generate music with multiple MIDI tracks. MusicVAE [13] introduces a hierarchical decoder into general VAE model to generate music note sequences with long-term structure. Due to the impressive results of Transformer in neural translation, Huang et al. modify this sequence model's relative attention mechanism and generate minutes of music clips with high long-range structural coherence [14]. In addition to the study of music structure, researchers also employ deep generative models to model music styles, such as producing jazz melodies through two LSTM networks [15], harmonizing a user-made melody in Bach's style [16]. Most of them are trained on the specific style dataset. The music generated from these models can only mimic the single style embodied in the training data.

Moreover, little attention has been paid to Chinese music generation with deep learning techniques, especially for modeling the music style of Chinese music, though some researchers utilize Seq2Seq model to create multi-track Chinese popular songs from scratch [17] or generate melody of Chinese popular songs with given lyrics [18]. The existing generation algorithms for Chinese traditional songs are mostly based on non-deep models such as Markov models [19], genetic algorithms [20]. These studies cannot break up the bottleneck in melody creation and style imitation.

Some latest work in the domain of music style transfer begins to generate music with mixed style or recombine music content and style. For example, Mao et al. propose an end to end generative modal to produce music with mixture of different classical composer styles [21]. Lu et al. study the deep style transfer between Bach chorales and Jazz [22]. Nakamura et al. complete melody style conversion among different music genres [23]. The above studies are based on the music data from different genres or composing periods. However, the regional style generation of Chinese folk songs studied here is modeling style within the same genre, which is more challenging.

3 Approach

3.1 Music Representation

The monophonic folk songs M can be represented as a sequence of note tokens, which is a combination of its pitch, interval and rhythm. Pitch and rhythm are essential information for music. The interval is an important indicator to distinguish the regional music feature, especially for Han Chinese folk songs[24]. The detail processing is described as below and shown in Fig. 1.

(1) **Pitch Sequence** P: Sequence of pitch tokens which consists of the pitch type presented in melody sequence. Rest note is assigned a special token.
(2) **Interval Sequence** I: Sequence of interval tokens derived from P. Each interval token is represented as a deviation between the next pitch and current pitch in step of semitone.

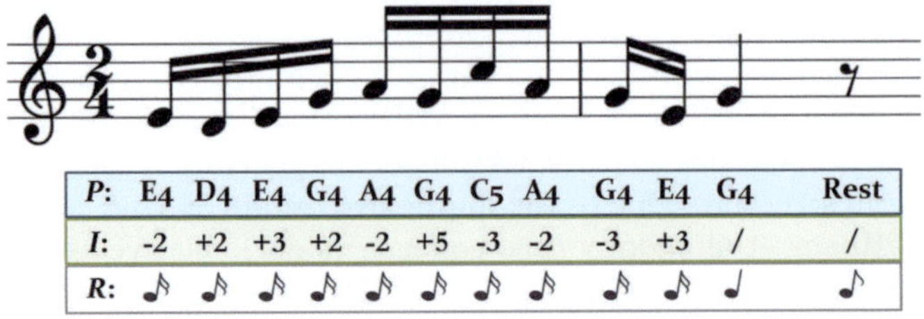

Fig. 1. Chinese folk songs representation including pitch sequence, interval sequence, rhythm sequence

(3) **Rhythm Sequence** R: Sequence of duration tokens comprised of the duration type presented in melody sequence.

3.2 Model

As mentioned in Section 1, the regional characteristics of Chinese folk songs are mainly reflected in their pitch patterns and rhythm patterns. In some areas, the regional characteristics of folk songs are more dependent on pitch feature, while the rhythm patterns in some areas are more distinctive. For example, in terms of pitch, folk songs in northern Shaanxi tend to use perfect forth, the Hunan folk songs often use the combination of major third and minor third [3], while Uighur folk songs employ the non-pentatonic scale. In terms of rhythm, Korean folk songs have their special rhythm system named *Jangdan*, while Mongolian *Long Song*s generally prefer long duration notes [25].

Inspired by the above observations, it is necessary to further refine the style of folk songs both in pitch and rhythm. Therefore, we propose a VAE-based model to separate pitch space and rhythm space, and further disentangle the music style and content space from pitch and rhythm space, respectively.

VAE and its Latent Space Division The VAE introduces a continuous latent variable z from a Gaussian prior $p_\theta(z)$, and then generates sequence x from the distribution $p_\theta(x|z)$ [26]. Concisely, a VAE includes an encoder $q_\phi(z|x)$, a decoder $p_\theta(x|z)$ and latent variable z. The loss function of VAE is

$$J(\phi,\theta) = -\mathbb{E}_{q_\phi(z|x)}[\log p_\theta(x|z)] + \beta KL(q_\phi(z|x)\|p_\theta(z)) \tag{1}$$

where the first term denotes reconstruction loss, and the second term refers to the Kullback-Leibler (KL) divergence, which is added to regularize the latent space. Weight β is a hyperparameter to balance the two loss terms. By setting $\beta < 1$, we can improve the generation quality of the model [27]. $p_\theta(z)$ is the prior and generally obeys the standard normal distribution, i.e., $p_\theta(z) = \mathcal{N}(0, I)$. The posterior approximation $q_\phi(z|x)$ is parameterized by encoder which is also

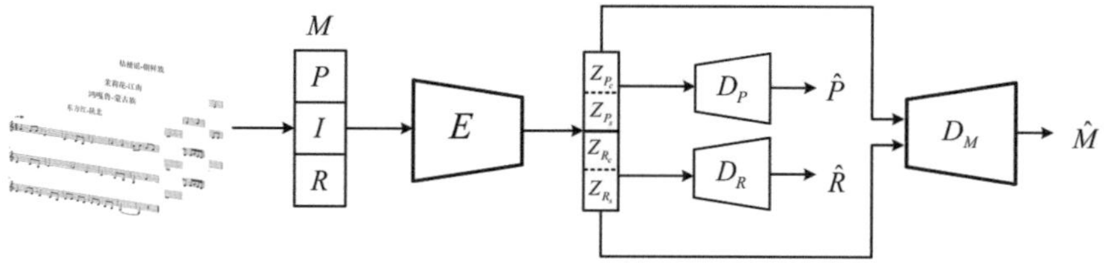

Fig. 2. Architecture of our model, it consists of melody encoder E, pitch decoder D_P, rhythm decoder D_R and melody decoder D_M

assumed to be Gaussian and reparameterization trick is used to acquire its mean and variance.

With the labeled data, we can disentangle the latent space of VAE in a way that different parts of the latent space correspond to different external attributes, which can enable the generation process in a more controllable way. In our case, we assume that the latent space can be firstly divided into two independent parts, i.e., *pitch variable* and *rhythm variable*. The pitch variable learns the pitch features of Chinese folk songs, while rhythm variable captures the rhythm patterns. Further, we assume both the pitch variable and rhythm variable consist of two independent parts, which refer to music *style variable* and music *content variable*, respectively.

Specifically, given a melody sequence $M = \{m_1, m_2, \cdots, m_n\}$ as the input sequence with n tokens (notes), where m_k denotes the feature combination of the corresponding pitch token p_k, interval sequence i_k and rhythm sequence r_k, we firstly encode M and obtain four latent variables from the linear transformation of the encoder's output. The four latent variables are pitch style variable Z_{P_s}, pitch content variable Z_{P_c}, rhythm style variable Z_{R_s} and rhythm content variable Z_{R_c}, respectively. Then, we concatenate Z_{P_s} and Z_{P_c} into the total pitch variable Z_P, which is used to predict the pitch sequence $\hat{P}$. The same operation is launched in rhythm variable to predict $\hat{R}$. Finally, all latent variables are concatenated to predict the total melody sequence $\hat{M}$. The architecture of our model is shown in Fig. 2.

Based the above assumption and operation, it is easy to extend the basic loss function:

$$J_{vae} = H(\hat{P}, P) + H(\hat{R}, R) + BCE(\hat{M}, M) + \beta KL_{total} \tag{2}$$

where $H(\cdot, \cdot)$ and $BCE(\cdot, \cdot)$ denote the cross entropy and binary cross entropy between prediction values and target values, respectively, and KL_{total} denotes the sum KL loss of the four latent variables.

Adversarial Training for Latent Spaces Disentanglement Here, we propose an adversarial training based method to conduct the disentanglement of pitch and rhythm, music style and content. The detail processing is shown in Fig. 3.

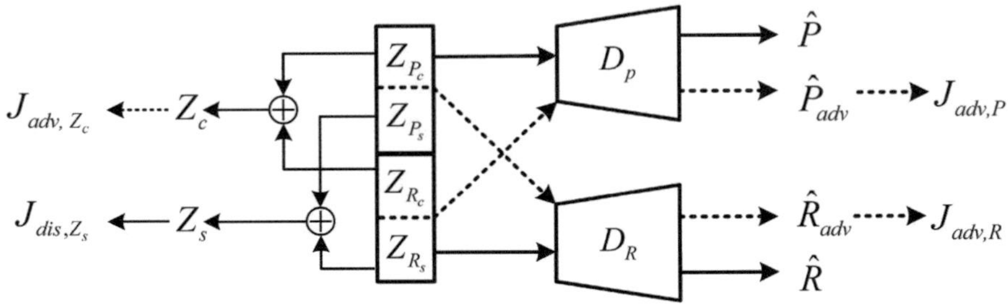

Fig. 3. Detail processing of latent spaces disentanglement. The dashed lines indicate the adversarial training parts

As shown in Fig. 2, we use two parallel decoders to reconstruct pitch sequence and rhythm sequence, respectively. Ideally, we expect the pitch variable Z_P and rhythm variable Z_R should be independent of each other. However, the pitch feature may be implicit in rhythm variables actually, vice versa, since the two variables are sampled from the same encoder output.

In order to separate the pitch and rhythm variable explicitly, the temporal supervision is employed in the separation of pitch and rhythm, which is similar to the work of disentangled representation for pitch and timbre [28]. Specifically, we feed the latent variable to the wrong decoder deliberately and force the decoder to predict nothing, i.e., all zero sequence, resulting in the following two loss terms based on cross entropy:

$$J_{adv,P} = -\Sigma[\mathbf{0} \cdot \log\hat{P}_{adv} + (1 - \mathbf{0}) \cdot \log(1 - \hat{P}_{adv})] \tag{3}$$

$$J_{adv,R} = -\Sigma[\mathbf{0} \cdot \log\hat{R}_{adv} + (1 - \mathbf{0}) \cdot \log(1 - \hat{R}_{adv})] \tag{4}$$

where $\mathbf{0}$ denotes all zero sequence, '$\cdot$' denotes the element-wise product.

For the disentanglement of music style and content, we firstly obtain the total music style variable Z_s and content variable Z_c:

$$Z_s = Z_{P_s} \oplus Z_{R_s}, Z_c = Z_{P_c} \oplus Z_{R_c} \tag{5}$$

where $\oplus$ means the concatenate operation.

Then two classifiers are defined to force the separation of style and content in the latent space using the regional information. The style classifier ensures the style variable is discriminative for regional label, while the adversary classifier force the content variable is not distinctive for regional label. For style classifier is trained with the cross entropy defined by

$$J_{dis,Z_s} = -\Sigma y \log p(y|Z_s) \tag{6}$$

where y denotes the ground truth, $p(y|Z_s)$ is the predicted probability distributions from style classifier.

For adversary classifier, we train it by maximizing the empirical entropy of the adversary classifier's prediction [29][30]. The training processing is divided into two steps. Firstly, the parameters of the adversary classifier are trained independently, i.e., the gradients of the classifier don't propagate back to VAE. Secondly, we compute the empirical entropy based on the output from adversary classifier as defined by

$$J_{adv,Z_c} = -\Sigma p(y|Z_c) \log p(y|Z_c) \tag{7}$$

where $p(y|Z_c)$ is the predicted probability distributions from adversary classifier.

In summary, the overall training objective of our model is the minimization the loss function defined by

$$J_{total} = J_{vae} + J_{adv,P} + J_{adv,R} + J_{dis,Z_s} - J_{adv,Z_c} \tag{8}$$

4 Experimental Results and Analysis

4.1 Datasets and Preprocessing

The lack of large-scale Chinese folk song datasets makes it impossible to apply deep learning methods for automatic generation and analysis of Chinese music. Therefore, we digitize more than 2000 Chinese folk songs in MIDI format from the record of *Chinese Folk Music Integration*[4]. These songs contain Han folk songs from Wu dialect district, Xiang dialect district[5] and northern Shaanxi, as well as three ethnic minority folk songs of Uygur in Xinjiang, Mongolian in Inner Mongolia and Korean in northeast China.

All melodies in datasets are transposed to C key. We use the Pretty-midi python toolkit [32] to process each MIDI file, and count the numbers of pitch

[4] *Chinese Folk Music Integration* is one of the major national cultural project leaded by the former Ministry of Culture, National Ethnic Affairs Commission and Chinese Musicians Association from 1984 to 2001. This set of book contains more than 40000 selected folk songs of different nationalities. The project website is http://www.cefla.org/project/book.

[5] According to the analysis of Han Chinese folk songs [24][31], the folk song style of each region is closely related to the local dialects. Therefore, we classify Han folk songs based on dialect divisions. Wu dialect district here mainly includes Southern Jiangsu, Northern Zhejiang and Shanghai. Xiang dialect district here mainly includes Yiyang, Changsha, Hengyang, Loudi and Shaoyang in Hunan province.

token, interval token and rhythm token as the feature dimension of the corresponding sequence, which are 40, 46 and 58, respectively. Then pitch sequence, interval sequence and rhythm sequence are extracted from raw notes sequence with the overlapping window of length 32 tokens and a hop-size of 1. Finally, we get 65508 ternary sequences in total. The regional labels of the token sequences drawn from the same song are consistent.

4.2 Experimental Setup

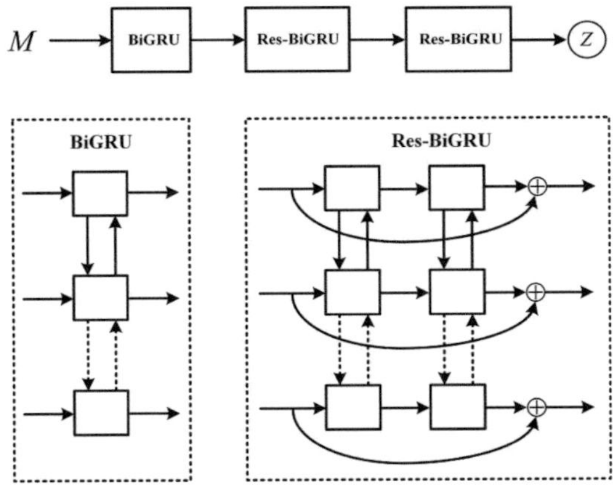

Fig. 4. Encoder with residual connections

In order to extract melody feature into latent space effectively, we employ a bidirectional GRU model with the residual connection [33] as encoder, which is illustrated in Fig. 4. The decoder is a normal two-layers GRU. All recurrent hidden size in this paper is 128. Both style classifier and adversary classifier are one-layer linear layer with Softmax function. The size of pitch style variable and rhythm style variable is set to 32, while the size of pitch content variable and rhythm content variable is 96. During training period, the KL term coefficient β increases from 0.0 to 0.15 linearly to alleviate the impact of posterior collapse.

Adam optimizer is employed with the initial learning rate of 0.01 for VAE training, and vanilla SGD optimizer with the initial learning rate of 0.005 for classifiers. All test models are trained for 30 epochs and the size of mini-batch is set to 50.

4.3 Evaluation and Results Analysis

To evaluate the generated music, we employ the following metrics from objective and subjective perspectives.

- **Reconstruction Accuracy**: We calculate the accuracy between the target notes sequence and reconstructed notes sequence on our test set to evaluate the music generation quality.
- **Style Recognition Accuracy**: We train a separate style evaluation classifier using the architecture in Fig. 4 to predict the regional style of the tunes that are generated using different latent variables. The classifier achieves a reasonable regional accuracy on the independent test set, which is up to 82.71%.
- **Human Evaluation**: As human should be the ultimate judge of creations, human evaluations are conducted to overcome the incoordinations between objective metrics and user studies. We invite three experts who are well educated and expertise in Chinese music. Each expert is asked to listen to the random selected five folk songs of each region on-site, and rate each song on a 5-point scale from 1 (very low) to 5 (very high) according to the following two criteria: a) *Musicality*: Does the song have a clear music pattern or structure? b) *Style Significance*: Does the songs' style match the given regional label?

Table 1. Results of automatic evaluations

Objectives	Reconstruction Accuracy	Style Recognition Accuracy
J_{vae}	0.7684	0.1726/0.1772/0.1814
$J_{vae}, J_{adv,P,R}$	0.7926	0.1835/0.1867/0.1901
$J_{vae}, J_{adv,P,R}, J_{adv,Z_c}$	0.7746	0.4797/0.4315/0.5107
$J_{vae}, J_{adv,P,R}, J_{dis,Z_s}$	**0.8079**	0.5774/0.5483/0.6025
J_{total}	0.7937	**0.6271/0.5648/0.6410**

Tab 1 shows all evaluation results of our models. The three values in the third column denote the accuracies derived from the following three kinds of latent variables: a) the concatenation of pitch style variable Z_{P_s} and a random variable sampled from standard normal distribution; b) the concatenation of rhythm style variable Z_{R_s} and the random variable; c) the concatenation of total style variable Z_s and the random variable. $J_{adv,P,R}$ denotes the sum of $J_{adv,P}$ and $J_{adv,R}$.

The model with J_{total} achieves the best results in style recognition accuracy and a sub-optimal result in reconstruction accuracy. The model without any constraints performs poorly on the two objective metrics. The addition of $J_{adv,P,R}$ improves the the reconstruction accuracy but fails to bring meaningful improvement to style classification. With the addition of either J_{adv,Z_c} or J_{dis,Z_s}, all the three recognition accuracies improve a lot, which indicates that the latent spaces are disentangled into style and content subspaces as expected. Moreover, only employing pitch style or rhythm style for style recognition can also obtain fair results, demonstrating the disentanglement of pitch and rhythm is effective.

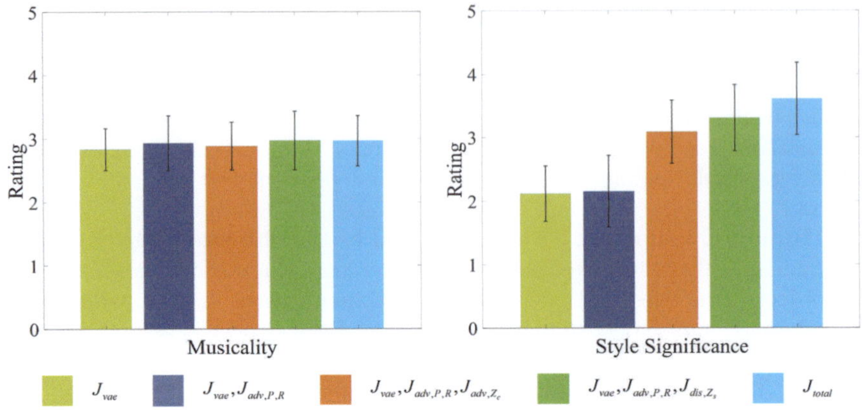

Fig. 5. Results of human evaluations including musicality and style significance. The heights of bars represent means of the ratings and the error bars represent the standard deviation

The result of human evaluations is shown in Fig. 5. In terms of musicality, all test models have similar performance, which demonstrates the addition of extra loss function has no negative impact on the generation quality of original VAE. Moreover, the model with total objectives J_{total} performs significantly better than other models in terms of style significance (two-tailed t-test, $p < 0.05$), which is consistent with the results in Tab 1.

Fig. 6 shows the t-SNE visualization[34] of our model with J_{total}. We can observe that music with different regional labels is noticeably separated in the pitch style space, rhythm style space and total style space, but looks chaos in content space.This further demonstrates the validity of our proposed methods to disentangle the pitch, rhythm, style and content.

Finally, we present several examples [6] of generating folk song with given regional labels with our methods in Fig. 7. As seen, we can create novel folk songs with dominated regional features such as long duration notes and large interval in Mongolian songs, the combination of major third and minor third in Hunan folk songs, and so on. However, there are still several failed examples. For instance, few generated songs repeat same melody pattern. More commonly, some songs don't show the correct regional feature, especially when the given regions belong to Han nationality areas. This may due to the fact that folk tunes in those regions share the same tonal system.

5 Conclusion

In this paper, we focus on how to capture the regional style of Chinese folk songs and generate novel folk songs with specific regional labels. We firstly col-

[6] Online Supplementary Material: https://csmt201986.github.io/mgvaeresults/.

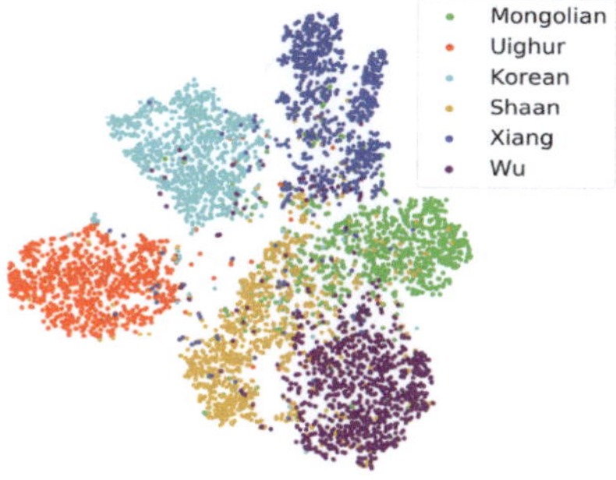

(a) Pitch style latent space

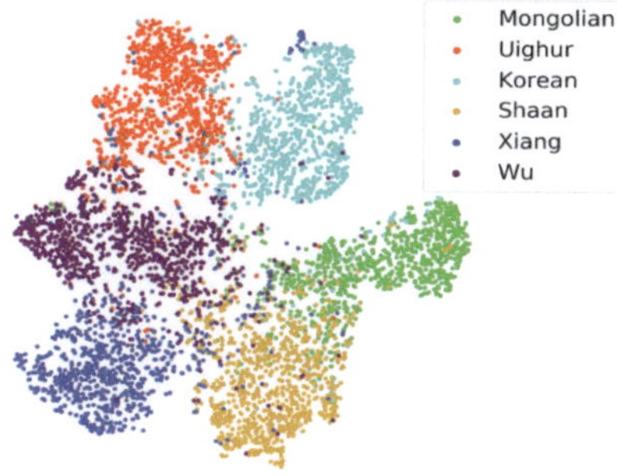

(b) Rhythm style latent space

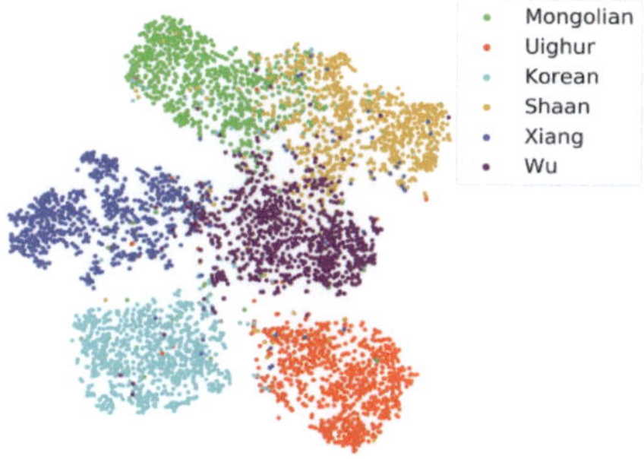

(c) Total style latent space

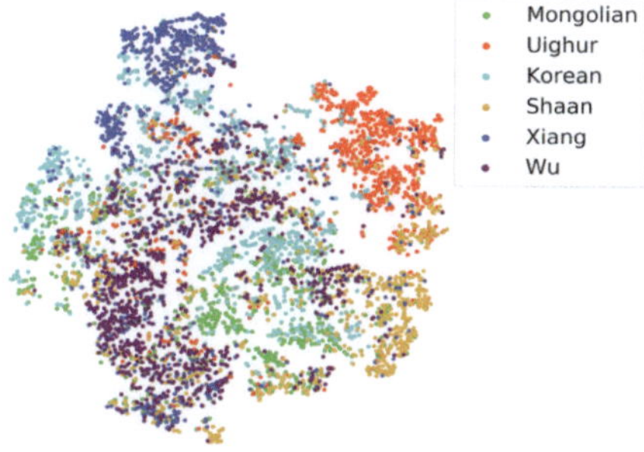

(d) Total content latent space

Fig. 6. t-SNE visualization of model with J_{total}

lect a database including more than 2000 Chinese folk songs for analysis and generation. Then, inspired by the observation of the regional characteristics in Chinese folk songs, a model named MG-VAE based on adversarial learning is proposed to disentangle the pitch variable, rhythm variable, style variable and content variable in the latent space of VAE. Three metrics containing automatic and subjective evaluation in our experiments are used to evaluate the proposed model. Finally, the experimental results and t-SNE visualization show that the disentanglement of the four variables is successful and our model is able to generate folk songs with controllable regional style. In the future, we plan to expand the proposed model to generate longer melody sequence using more powerful model like Transformers, and explore the evolution of tune families like *Mo Li Hua Diao*, *Chun Diao* among different regions.

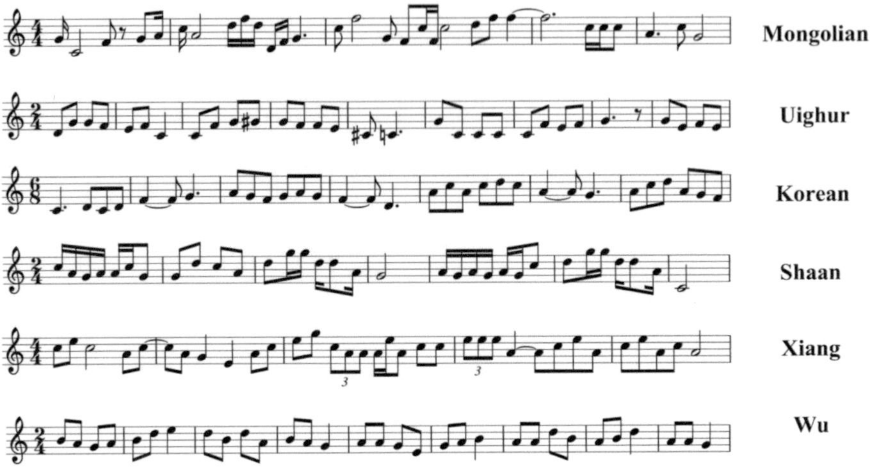

Fig. 7. Examples of folk songs generation given regional labels. In order to align each row, the scores of several regions are not completely displayed.

References

1. Dorien Herremans, Ching-Hua Chuan, and Elaine Chew. A functional taxonomy of music generation systems. *ACM Compututing Surveys*, 50(5):69:1–69:30, 2017.
2. Jean-Pierre Briot, Gaëtan Hadjeres, and François Pachet. Deep learning techniques for music generation - A survey. *CoRR*, abs/1709.01620, 2017.
3. Jing Miao and Jianzhong Qiao. A study of similar color area divisions in han folk songs. *Journal of Central Conservatory of Music*, 1(1):26–33, 1985.
4. Juan Li, Jing Luo, Jianhang Ding, Xi Zhao, and Xinyu Yang. Regional classification of chinese folk songs based on CRF model. *Multimedia Tools and Applications*, 78(9):11563–11584, 2019.
5. Shuqi Dai, Zheng Zhang, and Gus Xia. Music style transfer issues: A position paper. *CoRR*, abs/1803.06841, 2018.
6. Jianhua Guan. *The Contrast between Chinese and Western Music*. Nanjing Normal University Press, Nanjing, 2014.
7. Bob L. Sturm, João Felipe Santos, Oded Ben-Tal, and Iryna Korshunova. Music transcription modelling and composition using deep learning. *CoRR*, abs/1604.08723, 2016.
8. Dimos Makris, Maximos A. Kaliakatsos-Papakostas, Ioannis Karydis, and Katia Lida Kermanidis. Conditional neural sequence learners for generating drums' rhythms. *Neural Computing and Applications*, 31(6):1793–1804, 2019.
9. Sageev Oore, Ian Simon, Sander Dieleman, Douglas Eck, and Karen Simonyan. This time with feeling: Learning expressive musical performance. *Neural Computing and Applications*, pages 1–13, 2018.

10. Yujia Yan, Ethan Lustig, Joseph VanderStel, and Zhiyao Duan. Part-invariant model for music generation and harmonization. In *Proceedings of the 19th International Society for Music Information Retrieval Conference*, pages 204–210, Paris, France, 2018.
11. Li-Chia Yang, Szu-Yu Chou, and Yi-Hsuan Yang. Midinet: A convolutional generative adversarial network for symbolic-domain music generation. In *Proceedings of the 18th International Society for Music Information Retrieval Conference*, pages 324–331, Suzhou, China, 2017.
12. Hao-Wen Dong, Wen-Yi Hsiao, Li-Chia Yang, and Yi-Hsuan Yang. Musegan: Multi-track sequential generative adversarial networks for symbolic music generation and accompaniment. In *Proceedings of the Thirty-Second AAAI Conference on Artificial Intelligence*, pages 34–41, New Orleans, Louisiana, USA, 2018.
13. Adam Roberts, Jesse Engel, Colin Raffel, Curtis Hawthorne, and Douglas Eck. A hierarchical latent vector model for learning long-term structure in music. In *Proceedings of the 35th International Conference on Machine Learning*, pages 4361–4370, Stockholm, Sweden, 2018.
14. Cheng-Zhi Anna Huang, Ashish Vaswani, Jakob Uszkoreit, Noam Shazeer, Curtis Hawthorne, Andrew M. Dai, Matthew D. Hoffman, and Douglas Eck. An improved relative self-attention mechanism for transformer with application to music generation. *CoRR*, abs/1809.04281, 2018.
15. Daniel D. Johnson, Robert M. Keller, and Nicholas Weintraut. Learning to create jazz melodies using a product of experts. In *Proceedings of the Eighth International Conference on Computational Creativity*, pages 151–158, Atlanta, Georgia, USA, 2017.
16. Cheng-Zhi Anna Huang, Tim Cooijmans, Monica Dinculescu, and Adam Robertsand Curtis Hawthorne. Coconet: the ml model behind today's bach doodle, Accessed July 9, 2019.
17. Hongyuan Zhu, Qi Liu, Nicholas Jing Yuan, Chuan Qin, Jiawei Li, Kun Zhang, Guang Zhou, Furu Wei, Yuanchun Xu, and Enhong Chen. Xiaoice band: A melody and arrangement generation framework for pop music. In *Proceedings of the 24th ACM SIGKDD International Conference on Knowledge Discovery & Data Mining*, pages 2837–2846, London, UK, 2018.
18. Hangbo Bao, Shaohan Huang, Furu Wei, Lei Cui, Yu Wu, Chuanqi Tan, Songhao Piao, and Ming Zhou. Neural melody composition from lyrics. *CoRR*, abs/1809.04318, 2018.
19. Chih-Fang Huang, Yu-Shian Lian, Wei-Po Nien, and Wei-Hua Chieng. Analyzing the perception of chinese melodic imagery and its application to automated composition. *Multimedia Tools and Applications*, 75(13):7631–7654, 2016.
20. Xiaomei Zheng, Lei Wang, Dongyang Li, Lin Shen, Yanyuan Gao, Weian Guo, and Yushan Wang. Algorithm composition of chinese folk music based on swarm intelligence. *International Journal of Computing Science and Mathematics*, 8(5):437–446, 2017.
21. Huanru Henry Mao, Taylor Shin, and Garrison W. Cottrell. Deepj: Style-specific music generation. In *12th IEEE International Conference on Semantic Computing*, pages 377–382, Laguna Hills, CA, USA, 2018.
22. Wei Tsung Lu and Li Su. Transferring the style of homophonic music using recurrent neural networks and autoregressive model. In *Proceedings of the 19th International Society for Music Information Retrieval Conference*, pages 740–746, Paris, France, 2018.

23. Eita Nakamura, Kentaro Shibata, Ryo Nishikimi, and Kazuyoshi Yoshii. Unsupervised melody style conversion. In *IEEE International Conference on Acoustics, Speech and Signal Processing*, pages 196–200, Brighton, United Kingdom, 2019.
24. Kuo-Huang Han. Folk songs of the han chinese: Characteristics and classifications. *Asian Music*, 20(2):107–128, 1989.
25. Yaxiong Du. *An overview of Ethnic Minorities Folk Music in China*. Shanghai Conservatory of Music Press, Shanghai, 2014.
26. Diederik P. Kingma and Max Welling. Auto-encoding variational bayes. In *2nd International Conference on Learning Representations*, pages 1–14, Banff, AB, Canada, 2014.
27. Irina Higgins, Loïc Matthey, Arka Pal, Christopher Burgess, Xavier Glorot, Matthew Botvinick, Shakir Mohamed, and Alexander Lerchner. beta-vae: Learning basic visual concepts with a constrained variational framework. In *5th International Conference on Learning Representations*, pages 1–22, Toulon, France, 2017.
28. Yun-Ning Hung, Yi-An Chen, and Yi-Hsuan Yang. Learning disentangled representations for timber and pitch in music audio. *CoRR*, abs/1811.03271, 2018.
29. Zhenxin Fu, Xiaoye Tan, Nanyun Peng, Dongyan Zhao, and Rui Yan. Style transfer in text: Exploration and evaluation. In *Proceedings of the Thirty-Second AAAI Conference on Artificial Intelligence*, pages 663–670, New Orleans, Louisiana, USA, 2018.
30. Vineet John, Lili Mou, Hareesh Bahuleyan, and Olga Vechtomova. Disentangled representation learning for text style transfer. *CoRR*, abs/1808.04339, 2018.
31. Yaxiong Du. The music dialect area and its division of han chinese folk songs. *Journal of Central Conservatory of Music*, 1:14–16, 1993.
32. Colin Raffel and Daniel PW Ellis. Intuitive analysis, creation and manipulation of midi data with pretty_midi. In *15th International Society for Music Information Retrieval Conference Late Breaking and Demo Papers*, pages 84–93, Taipei, Taiwan, 2014.
33. Kaiming He, Xiangyu Zhang, Shaoqing Ren, and Jian Sun. Deep residual learning for image recognition. In *2016 IEEE Conference on Computer Vision and Pattern Recognition*, pages 770–778, Las Vegas, NV, USA, 2016.
34. Laurens van der Maaten and Geoffrey Hinton. Visualizing data using t-sne. *Journal of machine learning research*, 9(Nov):2579–2605, 2008.

The Study of Mapping Strategies Between the Excitators of the Single-Reed Woodwind and the Bowed String

Song Wang[1], Marcelo M. Wanderley[2], and Gary Scavone[1]

[1] Centre for Interdisciplinary Research in Music Media and Technology
[2] Schulich School of Music, McGill University, Montreal H3A 1E3

Abstract. Mapping is one of the most important components in digital musical instruments. There have been many works on "longitudinal" mapping strategies from the input device to the sound synthesizer. Such mapping can be considered longitudinal because the mapping direction is in line with the information transfer direction. However, less research focuses on "transversal" mappings among input devices or sound synthesizers. In this paper, a transversal mapping strategy is explored between the excitators of bowed strings and single-reed woodwinds which aims to allow more natural use of a given controller to play the sound of another family of instruments. A three-layer mapping structure, namely the playing layer, the mathematical layer, and the physical layer, is built. The mappings in different layers are generated based on the analogy of the mathematical models of two excitators in the mathematical layer. As a result, in the playing layer, the bowing force and the bowing speed of the string instrument are mapped to the lip force and the mouth pressure of a single-reed woodwind, respectively. In the physical layer, the string velocity and the friction force at the bowed point are mapped to the acoustic pressure and the volume velocity in the mouthpiece, respectively. Finally, a Yamaha WX5 wind controller is used to drive the digital waveguide string model. Two different mapping strategies between the lip force and the bowing force are tested and the results are discussed.

Keywords: Mapping, Excitator, Digital Musical Instrument, Single-reed instrument, Bowed string

1 Introduction

Mapping between the input device and the synthesis device is one of the keys to the digital musical instrument [1, 2]. As the link connecting the controller and the synthesizer, the mapping defines a pathway that the input information follows to flow toward the output, which builds an implicit bond between the player's gesture and the sound output. Traditionally, mappings are used to describe the relationship between two sets of parameters. Different mapping techniques have been introduced to make the instrument more controllable, such

H. Li et al. (eds.), *Proceedings of the 7th Conference on Sound and Music Technology (CSMT)*, Lecture Notes in Electrical Engineering 635,
https://doi.org/10.1007/978-981-15-2756-2_9

as neural networks [3–6], interpolation [7–9] and abstract parameters and multi-layered mapping [10, 11]. However, most of previous studies focus on the longitudinal mapping from the input device to the synthesis device with a typically unidirectional flow of information. In this paper, we propose to extend mappings to transversal mappings among input devices or synthesis devices.

Input devices specifically designed for a particular type of music may not work very well to control other families of instruments. The same issue happens to the synthesis device who may work well only when its controlled by a certain type of controllers. In contrast with longitudinal mappings, the transversal mappings build a link among the same category of devices, aiming to extend the applicability of the input device and increase the adaptivity of the synthesis device. A schematic comparison between two kinds of mappings is shown in Fig. 1.

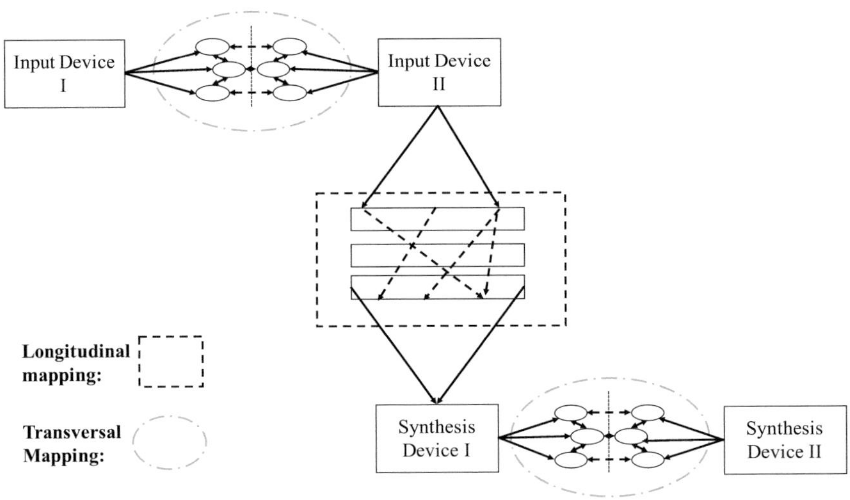

Fig. 1: A schematic view of the longitudinal and transversal mappings

In this paper, a special case of the transversal mapping is studied by exploring the mapping strategies between the excitators of single-reed woodwinds and bowed strings. Using the mapping strategies built in this paper, a wind controller can control the sound synthesizer of a bowed string, and vice versa. This paper is organized as follows: in Section 2, a three-layer mapping structure is proposed and applied to mapping between a bowed string and single-reed woodwinds. A preliminary experiment is conducted in the same Section, using a Yamaha WX5 wind controller to drive a digital waveguide bowed string model. Possible extensions and different interpretations of the mapping are discussed in Section 3. Finally, the conclusion is given in Section 4.

2 Mapping between the bowed strings and the single-reed woodwind

2.1 Three-layer mapping structure

The key to the transversal mapping between two different instruments is to map their playing parameters. However, because of the underlying complex physics, it is not straightforward to directly connect one set of playing parameters to another. In order to solve this problem, a three-layer mapping structure is proposed with three different mapping levels: the *playing layer* (playing parameters), the *mathematical layer* (mathematical models), and the *physical layer* (physical variables). As shown in Fig. 2, mapping on the mathematical layer is first based on the analogy of mathematical models between two instruments, which is shown as the solid line across instruments. Based on different mathematical models, the playing parameters and the physical variables are extracted as two separated layers. The mapping between the playing parameters or the physical variables of different instruments are determined by the role they play in their own mathematical model, shown as dashed lines across instruments. For example, both the bowing velocity in the string model and the mouth pressure in the single-reed model work as the energy source of the system, which makes it obvious to map them to each other.

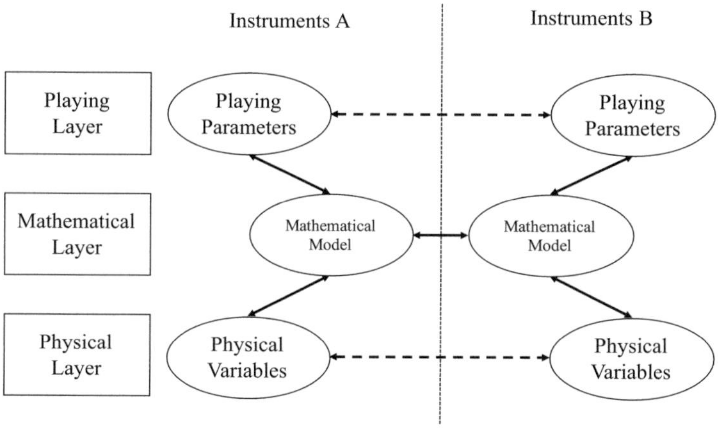

Fig. 2: Three-layer structure of mapping between two different musical instruments

In the next session, the generic mathematic model for self-sustained instruments will be introduced. The bowed string model, single-reed model and their mapping detail will then be discussed.

2.2 Mathematical models and mapping

Generic model Both bowed strings and single-reed woodwinds are classified as self-sustained musical instruments. As shown in Fig. 3, self-sustained musical instruments can generally be decomposed into a nonlinear excitator and a linear resonator. In the mathematical layer of the three-layer mapping structure, we take the excitator mathematical model as the mapping object instead of taking the whole instrument model. This is because the nonlinear excitator, sitting in between the player and the resonator, works as an information exchanger. It dominates sound generation by turning the continuous DC energy source provided by the player into an AC oscillation that excites the resonator. Hence, the excitator mathematical model involves both playing and physical variables, which makes it the best candidate to study mapping.

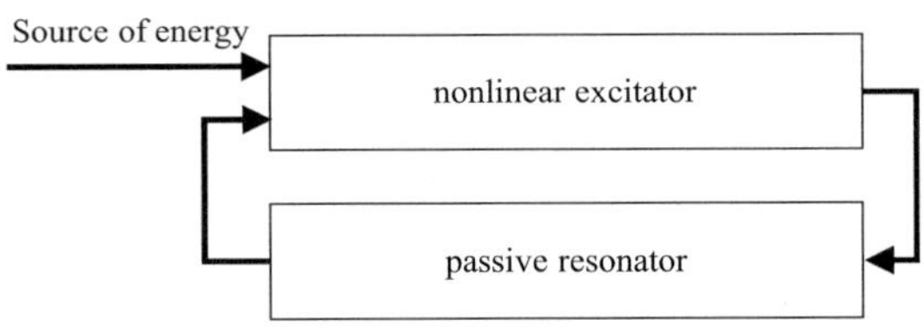

Fig. 3: Block diagram of a generic model for self-sustained instruments

Bowed string model The interaction of the bow and the string is usually considered as the stick-slip motion. The model proposed by Weinreich and Caussé [12] is used here. In this model, the contact area between the bow and the string is assumed as a point. The relationship between the friction force f and the velocity difference $\Delta v = v_b - v_s$ between the bow and the string is given by:

$$f = F_b(\frac{\Delta v}{v_0})[1 + (\frac{\Delta v}{v_0})^2]^{-1}, \tag{1}$$

where F_b is the bowing force, v_0 is a control parameter, v_b is the bowing velocity and v_s are the string velocity at the bowing point.

The friction force f as a function of the velocity difference Δv is shown in Fig. 4 with $F_b = 0.15\,N$ and $v_0 = 0.05\,m/s$.

Single-reed woodwind model For single-reed instruments, the nonlinearity is depicted by the relationship between the airflow rate in the mouthpiece u and the difference $\Delta p = p_m - p$ between the mouth pressure p_m and the mouthpiece pressure p. In this model, the nonlinear mechanism is assumed to be localized at the entrance of the mouthpiece chamber or the pipe resonator. Based on the

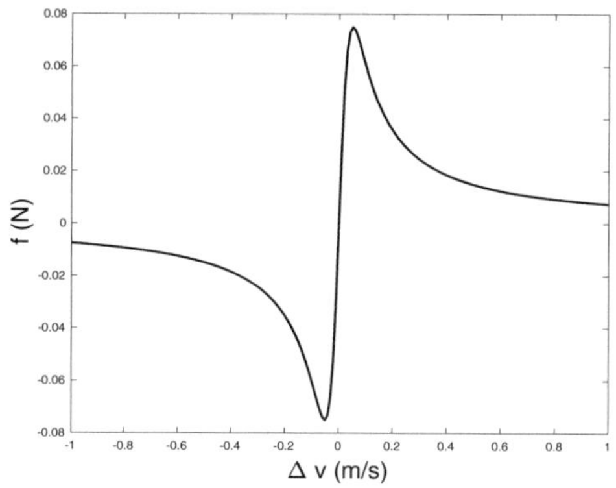

Fig. 4: Friction force in terms of the velocity difference

Bernoulli equation, the relationship is shown as:

$$u = S\sqrt{\frac{2|\Delta p|}{\rho}}\text{sign}(\Delta p), \tag{2}$$

where ρ is the density of the air and $S = wH$ is the cross-section area depending on the width of the reed channel w and the reed tip opening height H. However, when the pressure difference Δp is larger than a closing pressure p_M, the reed will touch the lay of the mouthpiece so that the channel will be closed ($H = 0$) and the airflow rate will be zero ($u = 0$). To keep things simple, the reed is modeled as an ideal spring without damping and inertia. The opening height is given by

$$H = \begin{cases} H_0 - \frac{\Delta p}{K}, & \text{for } p \leq p_M \\ 0, & \text{for } p > p_M \end{cases} \tag{3}$$

where K is the stiffness of the reed and $p_M = KH_0$.

Combining Eqs. 2 and 3, we get:

$$u = \begin{cases} w[H_0 - \frac{\Delta p}{K}]\sqrt{\frac{2|\Delta p|}{\rho}}\text{sign}(\Delta p), & \text{for } p \leq p_M \\ 0, & \text{for } p > p_M, \end{cases} \tag{4}$$

which is further simplified as

$$u = \begin{cases} u_A\left(1 - \frac{\Delta p}{p_M}\right)\sqrt{\frac{\Delta p}{p_M}}, & \text{for } p \leq p_M \\ 0, & \text{for } p > p_M, \end{cases} \tag{5}$$

where

$$u_A = w\sqrt{\frac{2KH_0^3}{\rho}}. \tag{6}$$

The nonlinear characteristic of this equation is shown in Fig. 5 with $w = 12$ mm, $H_0 = 0.3$ mm, $K = 188$ hPa/mm.

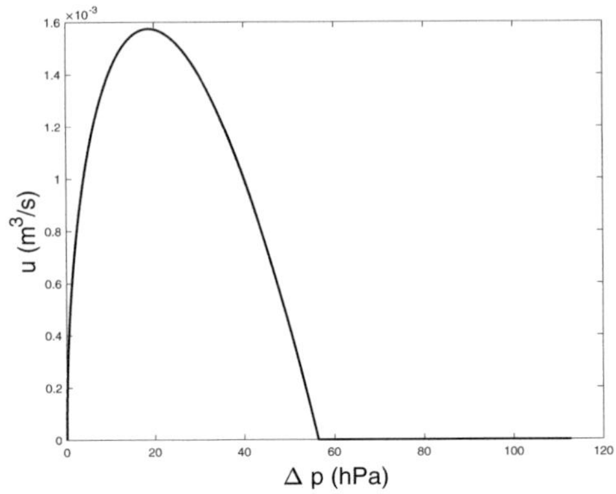

Fig. 5: The function of the flow rate in terms of the pressure difference

In previous research by Ollivier *et al.* [13], the embouchure flow u_A is taken as a playing parameter. However, u_A depends on multiple parameters like the tip opening H_0 and the reed stiffness K. Furthermore, such a parameter is not consciously and directly controlled by the player, and is unlikely to be taken as a playing parameter. Instead, the lip force F_{lip} acting on the reed is taken as the substitute playing parameter in this paper. The way to replace u_A by F_{lip} in the mathematical model is discussed below.

Based on the reed instrument mouthpiece model presented by Avanzini and van Walstijn [14, 15], the lip force influences the damping, stiffness, and also the rest opening height of the mouthpiece channel. Because the damping of the reed is not considered here, only the reed stiffness K and the reed rest opening height H_0 are taken into consideration as functions of lip force F_{lip}. Again, in order to keep things simple, instead of using the reed-mouthpiece-lip interaction model, linear functions built from the measurement data [16] are derived. Based on the data provided by Dalmont et al. [16], H_0 and $1/K$ as functions of F_{lip} are shown in Fig. 6.

A linear function is used to fit these relationships:

$$H_0 = -aF_{lip} + b, \tag{7}$$

$$\frac{1}{K} = -cF_{lip} + d, \tag{8}$$

where $a = 0.2426$, $b = 1.0614$, $c = 0.0024$ and $d = 0.0130$.

Substituting Eqs. 7 and 8 into Eq. 4, we get:

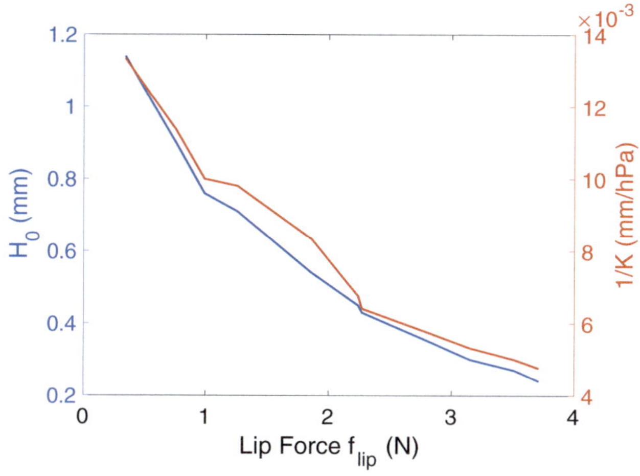

Fig. 6: H_0 and $1/K$ as functions of $F_{lip}h$

$$u = \begin{cases} [-F_{lip}(a + c\Delta p) + (b + d\Delta p)]\, w\sqrt{\dfrac{2|\Delta p|}{\rho}}\,\mathrm{sign}(\Delta p) & \text{for } p \leq p_M \\ 0 & \text{for } p > p_M \end{cases} \tag{9}$$

where F_{lip} is successfully extracted as an independent variable in the model.

Mapping The three-layer mapping between the bowed string and the single-reed woodwind excitations is shown in Table 1.

In the mathematical layer, the friction force equation and the Bernoulli equation are used to model the bowed string and the single-reed mouthpiece systems, respectively. Based on the analogy of these two sets of equations, the mappings in the physical layer and playing layer are built.

Table 1: Mapping between the excitations of bowed string and single-reed instruments

Mapping Layer	Bowed strings	Single-reed Instruments
physical layer	string velocity v at the bowed point friction force f at the bowed point	acoustic pressure p inside the mouthpiece volume velocity u in the mouthpiece
mathematical layer	friction force equation	Bernoulli equation
playing layer	bowing speed v_b bowing force F_b	pressure inside the mouth p_m lip force F_{lip}

Within the playing layer, the bowing speed v_b is mapped to the pressure in the mouth p_m as they generate energy sources in a similar way. In the bowed string instrument, a non-zero bowing speed leads to the friction between the bow and the string, which in turn provides the source of the velocity wave. Similarly, in the single-reed woodwind, the non-zero mouth pressure forces the air to pass through the reed channel and causes the reed to vibrate, which is the primary source of the pressure fluctuation. On top of the v_b-p_m pair, the lip force F_{lip} is directly mapped to the bowing force F_b. Despite the same physical meaning and unit of F_{lip} and F_b as forces, the lip force and the bowing force have the opposite influence on their physical models. In the bowed string model, a larger F_b results in a larger friction force, which increases the string velocity signal and leads to a louder sound. However, for a single-reed woodwind, as shown in the Eq. 9, a larger F_{lip} makes the u_A smaller, which lowers the sound amplitude. This effect is also explained by previous research and measurements by Almeida and colleagues [17]. Players will loosen their lip pressure for a louder sound or larger dynamics. From the clarinetist point of view, the increasing of the lip force should correspond to a decreasing of the bowing force. However, for a violinist or non-clarinetist, it is obvious to make a positive correlative mapping between these two different forces. Both positive and negative correlative mapping strategies are explored and discussed in the next section.

In the physical layer, the string velocity v at the bowing point is mapped to the acoustic pressure p inside the mouthpiece. Both of them can be used as the physical variables in the physical modeling of the resonator system [18]. In addition, the transverse friction force f is mapped to the volume velocity u inside the mouthpiece. These mappings also lead to a direct link between the instantaneous admittance $Y = v/f$ and impedance $Z = p/u$, which can be taken to describe the characteristics of the bowed string and the woodwind excitators, respectively.

2.3 Mapping experiments

In this section, a Yamaha WX5 wind controller is used to drive a bowed string digital waveguide (DWG) model. The MIDI information of the breath pressure and lip force is provided by the WX5 and taken as the input to the DWG model. The bowed string model is implemented in the Synthesis ToolKit (STK) [19] and packaged as a MAX/MSP object[3]. The interaction between the bow and the string is implemented by the "bow table" that is generated by Eq. 1.

Based on the mapping strategy discussed in the previous section, the mouth pressure is mapped to the bowing speed, and the lip force is mapped to the bowing force. As discussed earlier, the lip force can be mapped to the bowing force in two different ways so that two experiments are conducted, correspondingly.

In the first case, lip force is positively correlated to the bowing force. As the lip force increases, the sound gets louder, as well as the friction noise. In such a case, the control of the sound is easy and obvious. In the second experiment,

[3] See https://github.com/Cycling74/percolate.

the lip force is negatively correlated to the bowing force except for the initial state when lip force is zero. The zero lip force of the wind controller corresponds to the off-string bowing status in the bowed string model, which represents a zero bowing force. However, when the player starts to bite the mouthpiece, even very softly, the system will generate a large bowing force and make it very easy to generate sound. Such a mapping will require more and more pressure to generate noiseless sound with the increasing lip force The negatively correlated mapping makes the instrument harder to control, but in the meanwhile, makes it possible to play the bowed-string synthesizer as if one was playing a real clarinet. However, the embouchure is not exactly the same as the one used to play the real clarinet. This is mainly due to the linear mapping from the MIDI information of the WX5 to the bowing force, which could be improved in the future by introducing a nonlinear relationship.

3 Discussion

3.1 Mathematical model analogy

The key to the three-layer mapping between the bowed string and the single-reed woodwind is the mathematical model analogy. As mentioned by Ollivier et al. [13], the synthesis models have to be simple enough for comparison. Too many parameters of a model will make the analogy impossible even though there are several potential improvements to the current analogy and mapping.

First, a better lumped model of the single-reed woodwind could improve the mapping. As discussed above, the analogy between the mouthpiece and bowed string is based on the similarity between the pressure-flow and the stick-slip characteristics. In the present single-reed woodwind model, the volume velocity is set to zero when the pressure difference Δp is larger than the closing pressure p_M. In the paper by Walstijin and Avanzini [15], a different lumped model is proposed, which generates a more realistic u-Δp relationship by introducing the pressure-dependent stiffness. This model can make the two instrument characteristics more similar to each other as compared in Fig. 7.

Second, the hysteresis is an important effect that happens in both systems due to the characteristic of the reed and the string. As shown in Fig. 8, the two hysteresis effects captured by a complete coupled single-reed woodwind system [20] and the bowed-string model by Schumacher *et al.* [21] resembles each other. Such an analogy of hysteresis deserves a further study and test.

Third, in the current mapping experiment, the playing parameters of the single-reed instrument come directly from the wind controller. So the mapping is from the MIDI number of the single-reed instrument playing parameters to the physical value of the bowed-string playing parameters. However, more physically meaningful mappings can be made by considering the oscillation and extinction thresholds of playing parameters of both instruments. Such a threshold can be based on either modeling or measurement results.

Finally, friction noise is one of the essential elements that determines the naturalness of the sound. The noise is controlled by the bowing force and mouth

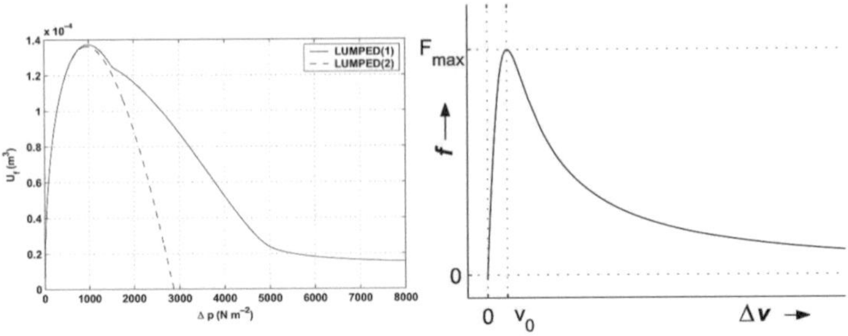

(a) (b)

Fig. 7: A basic characteristic of (a) the reed mouthpiece model (LUMPED(1) is the improved model) [15] and (b) the bowed string model [13].

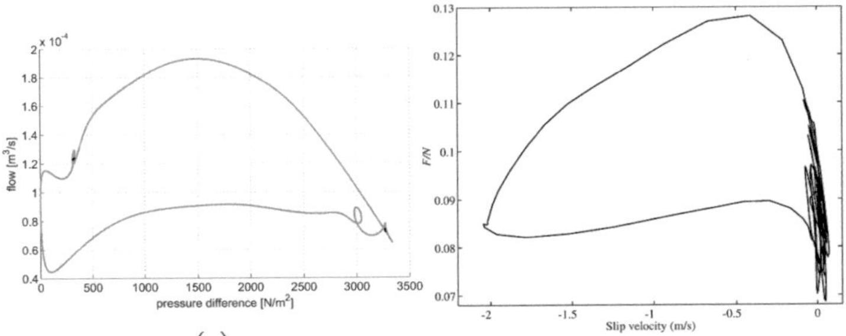

(a) (b)

Fig. 8: Hysteresis characteristics of (a) the reed mouthpiece model [20] and (b) the bowed string model [21].

pressure in the bowed string model and the single-reed model, respectively. However, the bowing force and the mouth pressure have no relationship with each other in the current mapping, so the noise of the bowed string might not be well controlled by a wind controller. A different mapping strategy is needed for better control of the noise, which may result in either a convergent or divergent mapping instead of the simple one-to-one mapping [22].

3.2 Discussion on the three-layer mapping structure

In this paper, the three-layer structure mapping is applied between the bowed string and the single-reed woodwind. In contrast to longitudinal mappings, this transversal mapping builds the relationship between the objects of the same category. However, both the input device and the synthesis device are involved in the three-layer mapping. So, this three-layer transversal mapping can also be interpreted as a part of the longitudinal mapping structure, being taken as an implicit mapping layer between the input device and the synthesis device.

Interestingly, the three-layer structure can be explained from both the system point of view and the functional point of view [23]. In terms of the system point of view, the output of the three-layer mapping is the mapping between two sets of parameters of different instruments, i.e., the playing parameter mapping in the playing layer and the physical variable mapping in the physical layer. In addition, these outer layer mappings are all based on the mathematical model analogy that can be considered as the mapping between functions from the functional point of view. In conclusion, the three-layer mapping between the bowed string and the single-reed woodwind has dual characteristics. It can be interpreted as both a transversal mapping and a longitudinal mapping. For the latter case, the mapping can be explained from both the system and functional point of views, which is attributed to the three-layer structure.

4 Conclusions

In this paper, a three-layer structure is proposed to study the mapping between the bowed string and single-reed woodwind. The mathematical model analogy in the mathematical layer provides a direct link between different instrument excitators. Based on the mapping in the mathematical layer, the playing parameters and physical variables are mapped in the playing layer and the physical layer, respectively. Based on such mapping, a Yamaha WX5 wind controller is used to drive the bowed string model. Two different strategies are tested, which provides a preliminary validation of the current mapping. When using positively correlated force mapping, the control is obvious and easy while counter-intuitive to wind performers. However, negatively correlated force mapping may provide more realistic experiences. Furthermore, the limitation and extensions of the current mapping strategies are discussed. Finally, the three-layer structure mapping is successfully interpreted from both the system point of view and the functional point of view.

References

1. Andy Hunt and Marcelo M Wanderley. Mapping performer parameters to synthesis engines. *Organised sound*, 7(02):97–108, 2002.
2. Eduardo Reck Miranda and Marcelo M Wanderley. *New digital musical instruments: control and interaction beyond the keyboard*, volume 21. AR Editions, Inc., Madison, WI, USA, 2006.
3. Matthew Lee and David Wessel. Connectionist models for real-time control of synthesis and compositional algorithms. In *Proceedings of the International Computer Music Conference*, pages 277–277, San Jose, 1992.
4. Sidney S Fels and Geoffrey E Hinton. Glove-talk: A neural network interface between a data-glove and a speech synthesizer. *IEEE transactions on Neural Networks*, 4(1):2–8, 1993.
5. Arshia Cont, Thierry Coduys, and Cyrille Henry. Real-time gesture mapping in pd environment using neural networks. In *Proceedings of the International Conference on New Interfaces for Musical Expression*, pages 39–42, London, 2004.

6. Chris Kiefer. Musical instrument mapping design with echo state networks. In *Proceedings of the International Conference on New Interfaces for Musical Expression*, pages 293–298, London, 2014.
7. Camille Goudeseune. Interpolated mappings for musical instruments. *Organised Sound*, 7(02):85–96, 2002.
8. Ross Bencina. The metasurface: applying natural neighbour interpolation to two-to-many mapping. In *Proceedings of the International Conference on New Interfaces for Musical Expression*, pages 101–104, Vancouver, 2005.
9. Martin Marier. Designing mappings for musical interfaces using preset interpolation. In *Proceedings of the International Conference on New Interfaces for Musical Expression*, Ann Arbor, 2012.
10. Marcelo M Wanderley, Norbert Schnell, and Joseph Rovan. Escher-modeling and performing composed instruments in real-time. In *SMC'98 Conference Proceedings. 1998 IEEE International Conference on Systems, Man, and Cybernetics (Cat. No. 98CH36218)*, volume 2, pages 1080–1084, San Diego, 1998.
11. Andy Hunt, Marcelo M Wanderley, and Ross Kirk. Towards a model for instrumental mapping in expert musical interaction. In *Proceedings of the International Computer Music Conference*, Berlin, 2000.
12. Gabriel Weinreich and Rene Causse. Elementary stability considerations for bowed-string motion. *The Journal of the Acoustical Society of America*, 89(2):887–895, 1991.
13. S Ollivier, J-P Dalmont, and J Kergomard. Idealized models of reed woodwinds. Part I: Analogy with the bowed string. *Acta acustica united with acustica*, 90(6):1192–1203, 2004.
14. Federico Avanzini and Maarten van Walstijn. Modelling the mechanical response of the reed-mouthpiece-lip system of a clarinet. Part I. A one-dimensional distributed model. *Acta Acustica united with Acustica*, 90(3):537–547, 2004.
15. Maarten van Walstijn and Federico Avanzini. Modelling the mechanical response of the reed-mouthpiece-lip system of a clarinet. Part II: A lumped model approximation. *Acta Acustica united with Acustica*, 93(3):435–446, 2007.
16. Jean-Pierre Dalmont and Cyrille Frappe. Oscillation and extinction thresholds of the clarinet: Comparison of analytical results and experiments. *The Journal of the Acoustical Society of America*, 122(2):1173–1179, 2007.
17. Andre Almeida, David George, John Smith, and Joe Wolfe. The clarinet: How blowing pressure, lip force, lip position and reed "hardness" affect pitch, sound level, and spectrum. *The Journal of the Acoustical Society of America*, 134(3):2247–2255, 2013.
18. Julius O Smith. Physical modeling using digital waveguides. *Computer music journal*, 16(4):74–91, 1992.
19. Perry R Cook and Gary Scavone. The synthesis toolkit (STK). In *Proceedings of the International Computer Music Conference*, pages 164–166, Beijing, 1999.
20. Vasileios Chatziioannou and Maarten van Walstijn. Estimation of clarinet reed parameters by inverse modelling. *Acta Acustica united with Acustica*, 98(4):629–639, 2012.
21. Robert T Schumacher, S Garoff, and J Woodhouse. Probing the physics of slip–stick friction using a bowed string. *The Journal of Adhesion*, 81(7-8):723–750, 2005.
22. Joseph Butch Rovan, Marcelo M Wanderley, Shlomo Dubnov, and Philippe Depalle. Instrumental gestural mapping strategies as expressivity determinants in computer music performance. In *Kansei, The Technology of Emotion. Proceedings of the AIMI International Workshop*, pages 68–73, Genova, 1997.

23. Doug Van Nort, Marcelo M Wanderley, and Philippe Depalle. Mapping control structures for sound synthesis: Functional and topological perspectives. *Computer Music Journal*, 38(3):6–22, 2014.

A Comparison of Melody Created by Artificial Intelligence and Human Based on Mathematical Model

Ziming Li, Shengchen Li
Institute of information photonics and optical communications, Beijing University of Posts and Telecommunications, Beijing, China, 100876

Abstract. There have been many achievements in using Artificial Intelligence (AI) to automatically compose melody. But there still need statistical methods to evaluate the degree of similarity between AI-generated melody and human composed melody. With N-gram models, this research compared the distribution of transitions of a combination of rhythm and pitch of melodies between melodies generated by AI (Magenta of Google) and the original British and American folk songs pieces used for training. The result shows that the AI system is good at modeling the common patterns of note transitions which appear more frequently in the data set of human-composed melodies than humans, and such patterns appear more frequently in the generated melodies compared with human-composed melody. The experiment shows that using the N-gram model to analyze the distribution characteristics of pitch and rhythm features of melody to distinguish between the AI-generated melody dataset and human composed melody dataset is a direction worthy of further research.

Keywords: Melody, Similarity, N-gram, Pitch, Rhythm.

1 Introduction

Machine learning based approaches has been proposed in automatic music generation. In particular, computational intelligence is widely used and has resulted in automatic melody generation [6]. However, there are few statistical evidences to evaluate the likelihood asserting whether a melody piece is composed by a machine or a human such that the evaluation of AI generated melody is often done by listening tests. The results of a listening test are affected by personal tastes of the audiences which makes the results of listening tests by different audiences incomparable. An objective method that evaluates the degree of similarity between AI-generated melody and human-composed melody can make different results comparable. The purpose of this research is to find an objective method that effectively evaluates degree of melody similarity by analyzing and comparing AI-generated melodies and human-composed melodies.

Currently, there are no objective definition of melody but researchers agree certain aspects of melody are important. For a specific problem such as comparing similarity of melody, the comparison of only a few features might be sufficient [12]. As people pay attention to rhythm and pitch changes for the perception of melody [13], this re-

H. Li et al. (eds.), *Proceedings of the 7th Conference on Sound and Music Technology (CSMT)*, Lecture Notes in Electrical Engineering 635,
https://doi.org/10.1007/978-981-15-2756-2_10

search evaluates the similarity between AI-generated melody and human composed melody in terms of both rhythm and pitch.

In order to investigate the changes in the rhythm and the pitch of the melody, a mathematical model is established. N-gram model is commonly used for exploring the probability relationship of state transitions hence N-gram models are used to represent the transition probability distribution as a feature in human-composed melodies. Then the same process repeats with AI-generated melodies. Finally, the degree of similarity between human-composed melody and AI-generated melody is evaluated by comparing the transition probability distribution in two types of melodies.

To avoid the impacts of melody style on the relationship between pitch and rhythm of melodies, this research selected British and American folk songs (including Jigs, Hornpipes, Morris, Reels and Ashover dances) as the dataset of human-composed of music to limit the style of generated music.

The remainder of the paper is organized as follows. Section 2 reviews the literature related to this research. Section 3 describes the melody simplification, the principle of N-gram model and the similarity evaluating algorithm. Section 4 presents experimental data, experiment design, and experiment results followed by a discussion and conclusion.

2 Literature review

Currently, most automatic composing systems use the pitch and the length of a note of human music as the primary input feature, neural networks and other machine learning algorithms are used to learn the pattern of human creation music [14][15]. Although measures from information theory has been used to compare music from Bach [17], most automated composing works evaluate the AI composed music through human evaluation, that is, let human beings listen to the music generated by AI and evaluate its similarity with human creation music [7][8][9][10][11]. This evaluate method is based on the subjective perception of human non-determinism and lacks a reliable statistical description. The purpose of this paper is to establish a mathematical model which uses theoretic mathematical results of timeline of melody to explore the similarities between AI-created music and human-generated music.

For the problem of music feature selection in symbolic music similarity research, there are many points of view for analyzing symbolic music with mathematical models [1][2]. Some works use string to represent music and the similarity is measured by calculating the distance of two strings representing two pieces of music [3]. Some methods use geometric objects to represent pieces of music, then calculate the geometric similarity of the two objects to evaluate the music similarity. There are some methods relying on music representations based on N-gram [5]. For similarity evaluation, N-gram can be used not only to calculate the similarity between two music pieces, but also be used to calculate the matching degree of music pieces with a specific music data set.

A research direction close to the direction of this research is to use the N-gram model for melody classification tasks. Pollastri, Emanuele, and Giuliano Simoncelli

[4] used HMMs of different length to classify classical music melodies composed by different composer which are represented as interval pitch and the result shows that memory of note is a fundamental method in classifying music by melodic content when timbre textures or harmony are not presented. Wei Chai and Barry Vercoe [16] used HMMs with different length to classify folk songs represented in different ways by countries. The results show that it is useful to use HMM to classify different folk songs according to the length and pitch features comparing with random guess (the classification accuracy is between 54% and 77%).

Inspired by above researches, this research uses the N-gram model to obtain the melody of British and American folk songs about the statistical features of pitch and length, which is compared with the AI-generated melody. Different from previous researches, the purpose of this research is to find a statistical evidence which can evaluate the degree of similarity between AI-generated melody and human-composed melody rather than classify them although the proposed experiment demonstrates a potential method classifying machine generated music.

3 The Method

3.1 N-gram

Suppose $\vec{\mathbf{X}}$ represents a sequence consisting of a series of samples ($\overrightarrow{x_1}$, $\overrightarrow{x_2}$, ..., $\overrightarrow{x_n}$) arranged in order, with n representing the length of the sample. Then we calculate the probability P(**X**) that the sample appears in the entire variable space, or denote P($\overrightarrow{x_1}$, $\overrightarrow{x_2}$, ..., $\overrightarrow{x_n}$), which can be decomposed according to the chain rule:

$$\mathrm{P}(\vec{\mathbf{X}}) = \mathrm{P}(\overrightarrow{x_1},\overrightarrow{x_2},\overrightarrow{x_3},\ldots,\overrightarrow{x_n}) = \mathrm{P}(\overrightarrow{x_1})\mathrm{P}(\overrightarrow{x_2}|\overrightarrow{x_1})\mathrm{P}(\overrightarrow{x_3}|\overrightarrow{x_1},\overrightarrow{x_2})\ldots\mathrm{P}(\overrightarrow{x_n}|\overrightarrow{x_1},\overrightarrow{x_2},\ldots,\overrightarrow{x_{n-1}} \quad (1)$$

The joint probability chain rule indicates that the occurrence of an event is related to each event that occurred before it. The N-gram model considers the context between n events. For example, when n=2 (bigram, 2-gram), the nth event $\overrightarrow{x_n}$ is related to the previous one events:

$$\mathrm{P}(\vec{\mathbf{X}}) = \mathrm{P}(\overrightarrow{x_1},\overrightarrow{x_2},\overrightarrow{x_3},\ldots,\overrightarrow{x_n}) = \mathrm{P}(\overrightarrow{x_1})\mathrm{P}(\overrightarrow{x_2}|\overrightarrow{x_1})\mathrm{P}(\overrightarrow{x_3}|\overrightarrow{x_2})\ldots\mathrm{P}(\overrightarrow{x_n}|\overrightarrow{x_{n-1}}) \quad (2)$$

In this research, note events are used as samples in the N-gram model to study transition relationships between note events.

3.2 Melody simplification

The data used in this research is a melody based on 88 piano keys. The pitch range between the highest note and the lowest note which appears in each melody is not uniform, but when human recognizes melody, people mainly focus on the interval relationship rather than the absolute pitch [16].So if the main tone of the tonality of the melody is used as a reference, the melodies with different tonalities may have the same transition pattern. So that a melody can be transposed while retaining the original note transition patterns. Unifying the pitch of the melody directly as absolute pitch (the actual pitch on the piano keys) may incorrectly classify the same transition pattern into different transition pattern. To avoid this potential problem, this research unifies the tonality of all the melodies in the dataset.

In this research, when extracting pitch feature of a melody piece, the root of the octave where the lowest pitch of the piece is located (the root note is determined by the tonality of the melody) is used as the lowest note, and the pitch of the melody piece is expressed as the relative pitch of the root. For example, if the lowest pitch of a melody piece in C major is C4 (261.63 Hz) in MIDI, then the relative pitch of the melody note is expressed as the number of semitones between its pitch and the root C4 (C4 is 0). In this research, all melody pieces in the training dataset are transposed to C major or C minor so the pitch of roots in every piece of melody is C. Tab. 1 is an example of the representation of the pitch of this method (the pitch of empty beats is unified into a special representation).

Table 1. An example of the representation of the pitch.

Lowest pitch in the melody piece	Representation of pitch of melody					
	C4	D4	E4	F4	G4	A4
C4 (261.63 Hz)	0	2	4	5	7	9
B3 (246.94 Hz)	12	14	16	17	19	21

The possibility of the duration of a note is theoretically infinite (infinitely long or infinitely short), so using absolute time as the input feature of the N-gram results in an output transition probability matrix that is sparse. To solve this problem, this research uses the continuous number of beats of the note event as the time feature. The note event length is expressed as an integer multiple of the basic time unit. The choice of basic time unit will be detailed in Section 4.1.

This research considers a melody as a two-dimensional vector consisting of the duration and pitch of a note event. The gap between two note events (the time period without sound) is also treated as a fixed pitch note event. With the pitch of a note event i represented as p_i, and the duration of a note event i is represented by d_i, and a melody progression can be expressed as $\overrightarrow{XM} = (\overrightarrow{xm_1}, \overrightarrow{xm_2}, \ldots, \overrightarrow{xm_n}) = [(p_1, d_1), (p_2, d_2), \ldots, (p_n, d_n)]$.

3.3 Similarity evaluation based on probabilities of transition probability matrix

3.3.1 Representations of feature space. After processing the melody data, in order to compare the AI-generated melody and the human-composed melody, a similarity evaluation method and standard are required. This research considers each different combination of $\overrightarrow{xm_l} = (p_l, d_l)$ as each different point in the two-dimensional vector space. When training an N-gram model, different points are treated as different states as the input of the N-gram model.

3.3.2 Similarity evaluation. First, we train the N-gram model with data from the existing folk dances created by humans whose note event transition probability matrix which is represented by *Ti*. Then we train the AI composition system with the same dataset as the N-gram model and generate a certain number and length of melodies. Whose transition probability matrix is represented by *Tm*. Next, the distribution of transition probability in *Tm* and *Tr* are compared. The similarity between each single AI-generated melody piece and human-composed melodies in the dataset is also evaluated. Finally, we discuss the similarity and difference between all the AI-generated melodies and human-composed melodies to test the performance of the similarity evaluation method used in this research.

The description of the measure for comparing similarity in this research is as follows: assume that, p_{xy}^{m} is the probability that note event *x* converts to note event *y* (a note event can be converted to the same note event) in *Tm*.

If the same note event transition x ⟶ y exists in *Tr* and the probability that note event *x* converts to note event *y* (a note event can be converted to the same note event) in *Tr* is represented as p_{xy}^{r}. The Euclidean distance D_{xy} between the melody piece to be evaluated and the human creation melody data regarding the note event transition is:

$$D_{xy} = \sqrt{(p_{xy}^{m} - p_{xy}^{r})^2} = |p_{xy}^{m} - p_{xy}^{r}| \tag{3}$$

If the note event transition x ⟶ y does not exist in *Tr*, the Euclidean distance D_{xy} between the melody piece to be evaluated and the human-composed melody data regarding the note event transition is:

$$D_{xy} = \sqrt{(p_{xy}^{m})^2} \tag{4}$$

Each D_{xy} of note event transition in *Tm* is calculated. Finally, the average Euclidean distance Dxa of all note events in the music piece is obtained (*n* in Formula (5) represents the total number of note events that appear in AI-generated melodies).

$$Dv = \left(\sum_{x=1}^{n} \sum_{y=1}^{n} D_{xy}\right) / n \tag{5}$$

Evaluating the similarity between the dataset of the melody piece to be evaluated and the dataset of human-created music by analyzing the distribution of Dv. The smaller the Dv, the closer the note transition probability of the evaluated melody is to the human-composed melody data set.

4 Experiments

4.1 Data

This research uses 594 MIDI files of human-composed British and American folk songs with 4/4 beats from (https://github.com/jukedeck/nottingham-dataset). The MIDI files provide two channels, one for melody and the other for the underlying chord progression. This research only uses the melody channel, and 534 MIDI files are used as the dataset to train the N-gram model. The eval/test set ratio is 8.9. The rest files are used to compare the similarity and differences with AI created melodies. Considering that most of the note events (97.2%) in the dataset are integer multiples of 0.25 beats. We set the basic time unit to 0.25 beat and the length of all note events (including empty note) in this research is expressed as integer multiple of 0.25 beats. The note event transitions containing a note whose duration is not an integer multiple of 0.25 beat are deleted (including both human-composed melody and AI-generated melody).

4.2 Experiment design

The process of this research is shown in Fig. 1. The N-gram model is trained by melody feature data obtained by the melody simplification of the MIDI data of the British and American folk songs melody, and the note event transition probability matrix *Tr* of the human-composed melody data is obtained. Then the AI-composition system trained by the data of the British and American folk songs melody generates a certain number of AI-generated melodies. Next, the *Tm* and *Dv* of each AI-generated melody and human-composed melody (test dataset) is calculated. Finally, we compare the *Dv* of human-composed melody data and AI-generated melody to evaluate the similarity and difference of human-composed melody data and AI-generated melody.

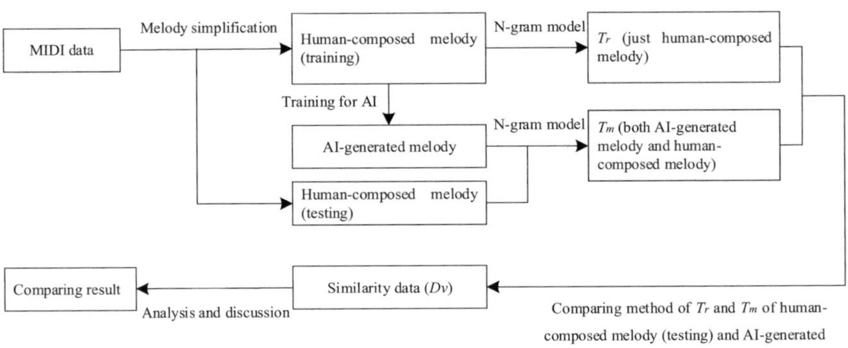

Fig. 1. The process of this research.

The creation and recognition of melody often involves the conversion relationship of multiple note events. The relationship of multiple note events is based on the conversion relationship of each pair of adjacent two note events. So, this research focuses on the state transitions of each pair of adjacent two note events in the melody. Therefore, this research sets the value of N for N-gram to 2. This research uses melody data to train a 2-gram model program written in MATLAB and get the note event transition probability matrix *Tr* of the melodies created by humans. Then we use same dataset to train the AI to generate 60 melodies. Next, we calculate the *Dv* of two groups of melody data: one is 60 human-composed melodies belong to the test dataset and another is 60 AI-generated melodies. Finally, we compare *Dv* of two groups of data to evaluating the similarity between human-composed melodies and AI-generated melodies.

The AI composing system used in this research is the Magenta launched by Google (https://github.com/tensorflow/magenta). Magenta extracts the melody data of the original data and inputs the neural network that generates the melody, and the network outputs the melody file in the MIDI format. A Recurrent Neural Network (RNN) named Attention_RNN in Magenta is used to generate melodies. Attention is a way that models can consider previous information without storing it in the state of RNN cells. When generating output for the current step, the previous n steps affect the output of the current step. We input the training dataset into Attention_RNN, and train after specifying the number of bars of generated melody, the number of training epochs, the number of attention steps and other parameters. The neural network outputs a melody MIDI file of a specified length.

The length of each melody is 32 bars (4/4 beats). The parameters of the network and training are as follows: batch-size: 128, RNN-layer-sizes: [128,128] (Two layers, with 128 neurons per layer), dropout probability: 0.5, learning-rate: 0.001, attention-step: 40, training epoch: 20000.

4.3 Result

The distributions of Dv of 60 AI-generated melodies and 60 human-composed melodies are shown in Fig. 2.

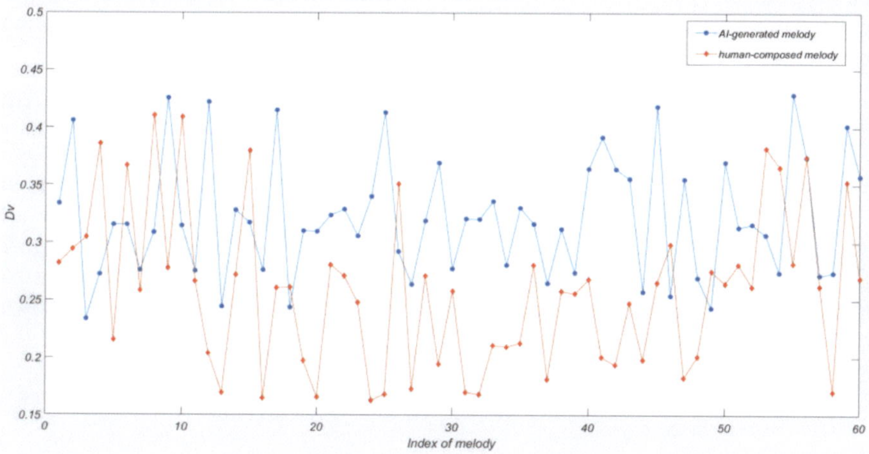

Fig. 2. The distributions of Dv of AI-generated melodies and human-composed melodies. Each hollow circle in the figure represents a melody. In order to facilitate the observation of the overall distribution, each discrete point is connected by a solid straight line.

Fig. 2 reflects some of the Dv distribution characteristics of human-composed melodies and AI-generated melodies:

1. There is an overlap interval between the Dv distributions of human-composed melodies and AI-generated melodies. The mean of Dv of human-composed melodies is lower than AI.

2. Dv fluctuations of different human-composed melodies are more obvious than different AI-generated melodies.

According to the information reflected in Fig. 2, we want to compare the Dv distribution differences between human-composed melodies and AI-generated melodies by statistics. So next, we calculated the average and variance of Dv of human-composed melodies and AI-created melodies. The results are shown in Tab. 2.

Table 2. The Average and Variance of Dv of human-composed and AI-generated melodies.

Composer	Average	Variance
Human	0.258	0.0046
AI	0.308	0.0028

It can be seen from Tab. 2 that the Dv variance of the AI-generated melody are smaller than the human-composed data. But the Dv average of the AI-generated melody are larger than the human-composed data. Dv can be used to distinguish the difference in this mathematical characteristic between the melody created by AI and the melody created by humans.

According to the results shown in Tab. 2, we speculate that the probability concentration degree of the probability matrix of human-composed melodies may be lower than the probability matrix of AI-generated melodies. In order to verify above speculations, an indicator *C-PRO* is proposed that measures the probability concentration degree of the probability matrix:

$$C - PRO = \left(\sum_{x=1}^{n}\sum_{y=1}^{n} p_{xy}\right)/n \tag{6}$$

the for where p_{xy} is the probability of note event x converts to note event y (x y) in a transition probability matrix. (n in Formula (6) represents the number of rows/columns of the n-square matrix). The higher the value of *C-PRO* is, the higher the average probability of the matrix is. We calculated the average and variance of *C-PRO* of human-composed melodies and AI-created melodies. The results are shown in Tab. 3.

Table 3. The Average and Variance of C-PRO of human-composed and AI-generated melodies.

	Composer	Average	Variance
1	Human	0.034	2.53×10^{-4}
2	AI	0.065	4.85×10^{-4}

It can be seen from Tab. 3 that the average and variance of *C-PRO* of the AI-generated melody are larger than the human-composed data. It is concluded that the note event transition probability of AI-generated melody is more concentrated but unstable than the note event transition probability of human-composed melody.

In order to verify the accuracy of the conclusion, we used a different training set and test set to do a 10-fold cross-validation (60 is a set of test sets), and found that the relationship of Dv of human-composed melody and Magenta-generated melody is consistent in 10 test results.

5 Discuss and conclusions

It can be seen from the comparison of the average and variance of Dv of the Magenta-generated melody and the human-composed melody that the distribution of the transition probability of the note events of the human-composed single melody is closer to the overall distribution of the note events transition probability of human-composed melody than the distribution of the transition probability of the note events of the Magenta-generated single melody in general. From the results shown in Tab. 3, the reason may be that Magenta is more concerned with the pattern of note event transitions with higher appearing frequency in the data set of human-composed melody, so that Magenta makes such patterns appear more frequently when generating melody. This makes the transition probability distribution of the note events of the Magenta-generated single melody more concentrated and monotonous than the human-composed single melody.

That is to say, if we use a N-gram model the Dv of an AI-generated melody data set and a human-composed melody dataset are obtained, it is possible to distinguish the two melody datasets according to the differences of the mean and variance of the Dv of these two datasets.

However, this research does not find a suitable statistical standard that can distinguish between a single AI-generated melody and a single human-composed melody, as well as distinguish AI-generated melody and human-composed melody with chord accompaniment, which will be the directions of further research in the future

References

1. Casey M A, Veltkamp R, Goto M, et al. Content-based music information retrieval: Current directions and future challenges[J]. Proceedings of the IEEE, 2008, 96(4): 668-696.
2. Wongsaroj C, Prompoon N, Surarerks A. A music similarity measure based on chord progression and song segmentation analysis[C]//2014 Fourth International Conference on Digital Information and Communication Technology and its Applications (DICTAP). IEEE, Bangkok, 2014: 158-163.
3. Van Kranenburg P, Volk A, Wiering F. A comparison between global and local features for computational classification of folk song melodies[J]. Journal of New Music Research, 2013, 42(1): 1-18.
4. Pollastri E, Simoncelli G. Classification of melodies by composer with hidden Markov models[C]//Proceedings First International Conference on WEB Delivering of Music. WEDELMUSIC 2001. IEEE, Florence, 2001: 88-95.

5. Ogihara M, Li T. N-Gram Chord Profiles for Composer Style Representation[C]//ISMIR. 2008, Pennsylvania: 671-676.
6. Liu C H, Ting C K. Computational intelligence in music composition: A survey[J]. IEEE Transactions on Emerging Topics in Computational Intelligence, 2016, 1(1): 2-15.
7. Liang F T, Gotham M, Johnson M, et al. Automatic Stylistic Composition of Bach Chorales with Deep LSTM[C]//ISMIR. 2017, **Suzhou**: 449-456.
8. Chu H, Urtasun R, Fidler S. Song from pi: A musically plausible network for pop music generation[J]. arXiv preprint arXiv:1611.03477, 2016.
9. Huang A, Wu R. Deep learning for music[J]. arXiv preprint arXiv:1606.04930, 2016.
10. Unehara M, Onisawa T. Composition of music using human evaluation[C]//10th IEEE International Conference on Fuzzy Systems.(Cat. No. 01CH37297). IEEE, Melbourne, 2001, 3: 1203-1206.
11. Maeda Y, Kajihara Y. Automatic generation method of twelve tone row for musical composition used genetic algorithm[C]//2009 IEEE International Conference on Fuzzy Systems. IEEE, Jeju, 2009: 963-968.
12. Volk A, Van Kranenburg P, Garbers J, et al. A Manual Annotation Method for Melodic Similarity and the Study of Melody Feature Sets[C]//ISMIR. 2008, Pennsylvania: 101-106.
13. Kim Y E, Chai W, Garcia R, et al. Analysis of a Contour-based Representation for Melody[C]//ISMIR. 2000, Massachusetts.
14. Dong H W, Hsiao W Y, Yang L C, et al. MuseGAN: Multi-track sequential generative adversarial networks for symbolic music generation and accompaniment[C]//Thirty-Second AAAI Conference on Artificial Intelligence. New Orleans, 2018.
15. Wu C L, Liu C H, Ting C K. A novel genetic algorithm considering measures and phrases for generating melody[C]//2014 IEEE Congress on Evolutionary Computation (CEC). IEEE, Beijing,2014,: 2101-2107.
16. Chai W, Vercoe B. Folk music classification using hidden Markov models[C]//Proceedings of international conference on artificial intelligence. sn, Seattle, 2001, 6(6.4).
17. Ren I Y. Using Shannon Entropy to evaluate automatic music generation systems: A Case Study of Bach's Chorales[J]. ECE Department, University of Rochester, 2015.

Part IV
Acoustic Signal Processing

Learning Multi-Resolution Representations for Acoustic Scene Classification via Neural Networks

Zijiang Yang[1], Kun Qian[2], Zhao Ren[1], Alice Baird[1], Zixing Zhang[3], and Björn Schuller[13]

[1] ZD.B Chair of Embedded Intelligence for Health Care & Wellbeing, Universität Augsburg, Germany, 86159
[2] Educational Physiology Laboratory, Graduate School of Education, The University of Tokyo, Japan, 113-0033
[3] Group on Language, Audio & Music, Imperial College London, The United Kingdom, SW7 2AZ

Abstract. This study investigates the performance of wavelet as well as conventional temporal and spectral features for acoustic scene classification, testing the effectiveness of both feature sets when combined with neural networks on acoustic scene classification. The TUT Acoustic Scenes 2017 Database is used in the evaluation of the system. The model with wavelet energy feature achieved 74.8 % and 60.2 % on development and evaluation set respectively, which is better than the model using temporal and spectral feature set (72.9 % and 59.4 %). Additionally, to optimise the generalisation and robustness of the models, a decision fusion method based on the posterior probability of each audio scene is used. Comparing with the baseline system of the Detection and Classification Acoustic Scenes and Events 2017 (DCASE 2017) challenge, the best decision fusion model achieves 79.2 % and 63.8 % on the development and evaluation sets, respectively, where both results significantly exceed the baseline system result of 74.8 % and 61.0 % (confirmed by one tailed z-test $p < 0.01$ and $p < 0.05$ respectively).

Keywords: Wavelets, Acoustic Scene Classification, Neural Networks, Machine Learning

1 Introduction

Features are the information contained in a data instance, e.g. pitch of an audio clip. In an effort to improve the features being extracted from a example researchers have developed an abundance of toolkits depending on specific needs. One popular toolkit to extract features of an audio instance is OPENSMILE [1]. OPENSMILE provides several feature extracting configurations, such as ComParE and eGeMAPS. The generated feature sets have been demonstrated to be effective for a range of tasks including autism and snore sound [2–4]. Other methods are also applied for acoustic classification tasks, including Mel Frequency Cepstral Coefficients (MFCCs) [5–8], Log Mel-band Energies [9], etc.

H. Li et al. (eds.), *Proceedings of the 7th Conference on Sound and Music Technology (CSMT)*, Lecture Notes in Electrical Engineering 635,
https://doi.org/10.1007/978-981-15-2756-2_11

Wavelet transformation (WT) is able to analyse both stationary and non-stationary signals whereas Fourier transformation has limited performance on non-stationary signal analysis. Moreover, WT has the ability to conduct multi-resolution analysis for signals, which standard Fourier transformation cannot [10]. Therefore, the performance of WT has been competitive on many tasks such as power quality classification, obstructive sleep apnea and auditory scene classification [11–13]. For instance, Qian et al. [2, 3] showed that applying wavelet features for the task of snore sound classification can achieve impressive results. Besides, Qian et al. [14] investigated wavelet features on acoustic classification task, where Support Vector Machine (SVM) and Gated Recurrent Neural Network (GRNN) were chosen as the classifiers. The result showed that wavelet energy features (WEF) with SVM achieved up to 77.8 % [14], which outperformed the baseline result (74.8 %) [9]. Qian's study showed that Wavelet features can reach competitive performances for the acoustic classification task.

Acoustic scene classification (ASC) is attracting researchers from multiple disciplines now [9, 7, 8, 15, 14, 16, 17]. The organisers of the IEEE Detection and Classification of Acoustic Scenes and Events Challenge (DCASE) provided a well-developed database [5] for researchers specifically. In DCASE 2017, the baseline applied a multilayer perceptron (MLP) as the classifier with log mel-band energies as features [9].

In the field of ASC, there have been competitive results found using SVM as a classifier. For example, Geiger et al. [8] has introduced SVM for their ASC submission to DCASE Challenge 2013. However, today most works choose neural network classifiers [14, 16, 17, 15], since researchers consistently show competitive results. For instance, the organiser of DCASE 2017 applied MLP in their baseline system. The setup of the system applying MLP, consisted of two dense layers (50 hidden units) and a 20 % dropout, as the classifier. Their MLP works for up to 200 epochs with a learning rate of 0.001. The features they chose are log mel-band energies, which are calculated at 40 ms per frame with a 50 % overlap, and the feature vector includes the context of 5 frames. Finally, they achieved good performance on both the development and evaluation set, 74.8 % and 61.0 % respectively [9].

Paticipants from DCASE 2017 used a neural network with a more complex structure. Valenti et al. [15] achieved 79.0 % and 86.2 % on the development and evaluation sets, which is 6.4 % and 9.0 % more than the baseline result of DCASE 2016, through using Convolutional Neural Networks (CNNs) and log-mel spectrogram features. Similarly, Ren et al. [16] applied a GRNN with spectrograms and scalograms features extracted achieving a 80.9 % accuracy.

Deep Neural Networks (DNNs) are now a popular extension of the neural networks. DNNs have succeeded in many fields of machine learning [18, 19]. In DCASE 2017, DNNs were applied by some participating teams. For example, Jung et al. [20] designed four systems with DNN structures and the best achieved an impressive accuracy of 95.9 % on the development set and 70.6 % on the evaluation set. Mun et al. [17] applied Generative Adversarial Networks (GANs) to generate new features and used SVM hyperplanes to select the most effective features from them. Then, a Fully Connected Neural Network (FCNN) and a SVM classifier were trained with the new database. This method highly improved the performance, and the late fusion of FCNN and SVM reached 87.1 % accuracy. Therefore, neural networks are a proven and effective solution in the field of ASC. In this paper, the information includes OPENSMILE [1] and wavelet features extracted from the DCASE data, with a deep MLP classifier. After investigating the performance on these two feature sets individually, a decision fusion based on probability is used at the end of the experiment to find out the best performance. The whole framework of this study is shown in Figure 1.

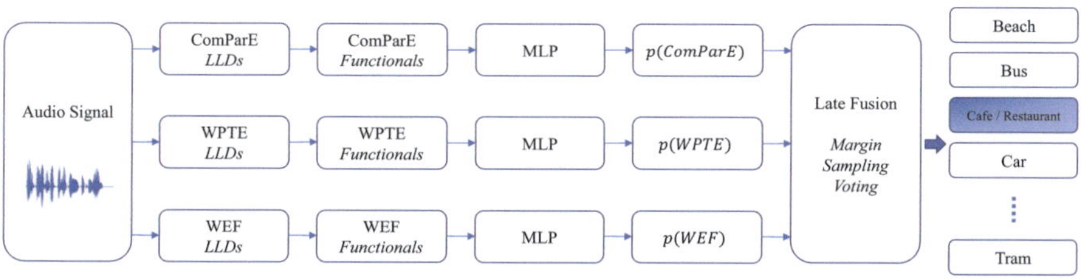

Fig. 1: Diagram of the Whole Framework of the Present Study

The rest of this paper will be organised as follows: An introduction to the feature sets will be shown in Section 2. Section 3 gives the description of the proposed methods. Experimental results are presented in Section 4 before a conclusion is made in Section 5.

2 Feature Sets

In this paper, two types of features are utilised, i. e., temporal and spectral features (extracted by OPENSMILE toolkit [1]), and wavelet features. As mentioned in Section 1, COMPARE feature set is chosen from this toolkit for its effectiveness and robustness. Most of the features in COMPARE feature set is Fourier transformation based but no WT included [21]. There are 55 spectral low-level descriptors (LLDs), 6 voicing related LLDs and 4 energy related LLDs. In total, 6 373 features are extracted from audio files [21]. Description is given in Table 1, where RASTA represents Relative Spectral Transform, HNR represents Harmonics to Noise Ratio and RMS represents Root Mean Square [14][21].

For wavelet features, two methods of WT, wavelet packet transformation (WPT) [22] and discrete wavelet transformation (DWT) [10], are investigated in this study. The features based on WPT can be expressed as [23, 24]:

$$\mathbf{E}_{\mathbf{S}_{j,m}} = \log \sqrt{\frac{\sum_{n=1}^{N_{j,m}} (\mathbf{w}_{j,m,n})^2}{N_{j,m}}}, \tag{1}$$

Table 1: ComParE Feature Set Description [14, 21]

Spectral LLDs
MFCC 1–14 Psychoacoustic sharpness, harmonicity RASTA-filt. aud. spect. bds. 1–26 (0–8 kHz) Spectral energy 250–650 Hz, 1 k–4 kHz Spectral flux, centroid, entropy, slope Spectral Roll-Off Pt. 0.25, 0.5, 0.75, 0.9 Spectral variance, skewness, kurtosis
Voicing Related LLDs
F_0 (SHS and Viterbi smoothing) Probability of voicing Log HNR, jitter (local and δ), shimmer (local)
Energy Related LLDs
RMS energy, zero-crossing rate Sum of auditory spectrum Sum of RASTA-filtered auditory spectrum

where $\mathbf{E}$ is the bank filter energy. $N_{j,m}$ means the number of coefficients at the j-th level in the m-th subband. The coefficient which is calculated from the subspace $\mathbf{S}_{j,m}$ is referred as to $\mathbf{w}_{j,m,n}$. m is set as all the integer from 0 to $2^j - 1$. In this research, the maximum decomposition level is applied to 7 [14]. Therefore, $2^{7+1} - 1 = 255$ LLDs are extracted. Then four statistical functionals are applied to those LLDs and $255 \times 4 = 1020$ features, referred as to wavelet packet transform energy (WPTE), are generated [14]. The DWT-based features are defined as [23]:

$$\widetilde{\mathbf{E}}_{\mathbf{S}_j} = \frac{(\mathbf{w}_j)^2}{\sum_{j=1}^{\mathbf{J_{max}}} (\mathbf{w}_j)^2} \times 100, \tag{2}$$

where $\mathbf{w}_j$ represents the coefficient calculated at the j-th level in DWT. After calculating mean, variance, waveform length and entropy, there are $4 \times (7+1) = 32$ LLDs are generated. Similarly to WPT, $32 \times 4 = 128$ features, referred as to wavelet transform energy (WTE), are extracted by DWT. The concatenation of WPTE and WTE is referred as to WEF, which has $1020 + 128 = 1148$ features [14].

3 Methodology

3.1 Database

In this paper, TUT Acoustic Scenes 2017 Database [9] were chosen as the database, since this database is professionally recorded, reliably partitioned, and can be easily accessed from the DCASE 2017 website [4]. According to the authors of TUT Acoustic Scenes 2016 Database [5] and 2017 Database [9], all audio files were professionally recorded and annotated. There are 15 different scenes in this database: *beach*, *bus*, *cafe/restaurant*, *car*, *city centre*, *forest path*, *grocery store*, *home*, *library*, *metro station*, *office*, *park*, *residential area*, *train*, and *tram*. Every recording has a average length of 3-5 minutes; then, all recordings are split into a length of 10 seconds and are considered independently. There are 312 clips for every class, 4 680 in total. [5]. In addition, comparing with TUT Acoustic Scenes 2016 Database, the 2017 version added 1 620 newly recorded clips (108 per class) as evaluation set [9].

In the first baseline experiment, the organisers divided the development set into four parts according to the following rule. They took one part as test set and the other samples were taken as the training set at random. All the 4 parts were taken as test set sequentially, therefore there were 4 sub-experiments or 4 folds. The average of all results was regarded as the accuracy for development. The organisers provide the meta data, which contains the rule for dividing samples, for researchers who want to challenge their baseline [5, 9].

However, the experiment on the evaluation set is simpler than above. The whole development set was taken as training set and the new 1 620 samples were included in the testing set [9]. The experiment in this paper applied the same setup with the baseline result provided, to make a clear objective comparison.

3.2 Classifier

An MLP is one of the classic models of a Neural Networks and has been applied in many tasks [9, 25, 26]. An MLP includes three main parts: an input layer, hidden layers and an output layer. Every layer has several neurons, or nodes, and they only interact with the neurons in adjacent layers. This means every neuron has no connection within the same layer [27]. Adjusting several hyperparameters of the MLP, such as the learning rate, the drop out rate and epochs, can directly influence the adjustment of neurons in the process, as well as influence the performance of the MLP for finding out the best configuration for a certain task.

3.3 Decision Fusion

All three feature sets, ComParE, WPTE and WEF, are used to train a model independently. Additionally, to improve the performance of the system, a decision fusion is applied. In the general prediction process, the model will generate a list, which includes the probabilities of every class in the classification task. Then the class which has the highest probability will be determined as the prediction of this sample. Moreover, there is additional operation if decision fusion is applied. After all three models (ComParE, WPTE and WEF) finish the prediction on the same sample, the model which has the largest difference between

[4] http://www.cs.tut.fi/sgn/arg/dcase2017/

the highest and the second highest probabilities in the list, also known as margin sampling value, will be determined as the best model [28]. Finally, the decision made by the best model will be the prediction of this decision fusion for the sample, as shown in Figure 1.

4 Experimental Results

4.1 Experimental Setup

In ASC processing, the audio file will be sent to the ASC system as input, and the output is the label which the classifier predicts to be the scene. ComParE features were extracted by the openSMILE toolkit [1] and two wavelet features, WPTE and WEF, were generated by the Wavelet Toolbox of Matlab [5]. After feature extraction, all features are standardised with zero mean and unit variance. The software used for building the MLP model is TensorFlow [29]. A two-layer MLP with all three kinds of features (ComParE, WPTE and WEF) are applied in the experiments, respectively. The final decision fusion, which is mentioned in Section 3.3, is used to improve the result after the best combination of hyperparameters is determined.

4.2 Results and Discussion

Firstly, the best configuration for every feature is determined by an abundance of experiments. All three features perform best in the two-layer MLP model. The three-layer and the four-layer model also have been tested on the development set, then both of them were abandoned according to the result. The best three-layer model achieved 72.3 % and the best four-layer model achieved only 17.9 %. The result showed that the best model is still two-layer model. All results of the two-layer model on the development set and the evaluation set are shown in Table 2 and 3.

Table 2: Results of All Models on the Development Set

Model	Fold 1	Fold 2	Fold 3	Fold 4	Mean
ComParE (A)	71.5	74.3	69.7	76.2	72.9
WPTE (B)	69.9	63.4	67.1	72.6	68.2
WEF (C)	75.2	76.8	73.5	73.8	74.8
A + B	75.6	73.4	75.6	79.0	75.9
A + C	78.9	80.3	78.0	79.6	**79.2**
B + C	74.9	72.6	72.7	74.6	73.7
A + B + C	77.5	76.9	77.9	79.5	77.9
Baseline					74.8

[5] https://www.mathworks.com/products/matlab.html

Table 3: Results of All Models on the Evaluation Set

Model	Accuracy
ComParE (A)	59.4
WPTE (B)	55.5
WEF (C)	60.2
A + B	62.7
A + C	**63.8**
B + C	60.9
A + B + C	63.5
Baseline	61.0

As given in Table 2 and 3, WEF achieves 74.8 % and 60.2 % of accuracy on the development and evaluation set, respectively. And it is the best of these three models. Besides, WEF is slightly better than ComParE, which achieves 72.9 % and 59.4 %. However, the baseline system result is still better than any of the three models that utilised only one kind of feature without decision fusion on the development set. For decision fusion models, all of these four models outweigh the baseline system, especially for ComParE + WEF, which achieves 79.2 % of accuracy. Moreover, the ComParE + WEF model is significantly better than the baseline system, according to a z-test [30] result ($p < 0.01$).

Additionally, only three models outweigh the baseline system on the evaluation set, and these models are: ComParE + WPTE with 62.7 %, ComParE + WEF with 63.8 % and ComParE + WPTE + WEF with 63.5 % of accuracy. The best model is ComParE + WEF, and it also significantly better than the baseline ($p < 0.05$). Although this result is not as good as the winner of DCASE 2017 Challenge, which reached 87.1 % accuracy on the evaluation set [17], there is still great potential here.

On both the development and the evaluation sets, models without the decision fusion, or in other words, models that use only one kind of features, perform worse than the baseline, except WEF on the development set which achieves the same accuracy as the baseline system. However, it is not significantly demonstrated that WEF is a better feature set than the ComParE feature set, for this task, due to the fact that the z test shows $p > 0.1$. This fact demonstrates that WEF and ComParE features have similar performance on ASC tasks, even though the combination of them succeeds on high accuracy in the task. In other words, using both of them with MLP in ASC tasks is a good idea in future researches. Also, comparing those results, applying decision fusion is able to improve the performance significantly and is worth further exploration.

To help the discussion on the results, the confusion matrix of four models (ComParE, WPTE, WEF and ComParE + WEF) on the evaluation set are given in Figure 2. The value shown in this matrix represent numbers of predicted instances. All of these four models have bad performance on *bus*, *library* and *park*. For instance, the model with ComParE features classifies only 31 of

108 *library* instances correctly, even though it is the best model of all fours for classifying class *library*. There are more information in these figures. For class *cafe/restaurant*, the model with WEF recognises 55 of 108 correctly. However, ComParE model does a better work and gets 92 of 108. Similar results happens on classifying *metro station*, where WEF model achieves 91/108 but ComParE gets only 50 correctly. These results demonstrate that every model has its own strengths and weaknesses on classification, and this fact can be used to improve the performance in the future, by adjusting weights in the decision fusion according to the strength of the model on classifying certain classes.

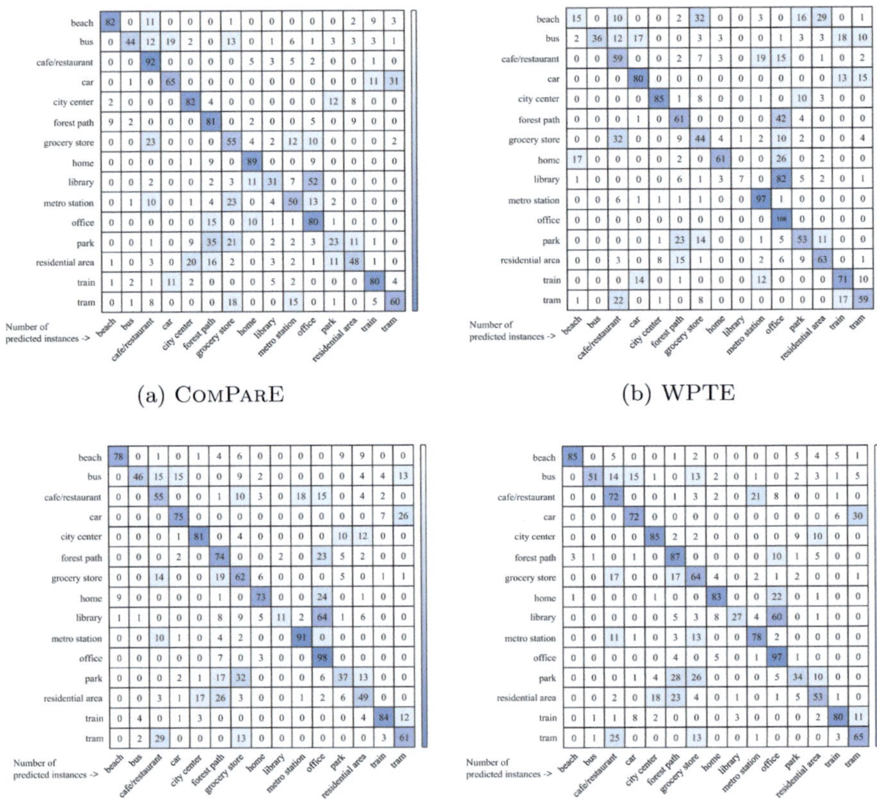

(a) ComParE (b) WPTE

(c) WEF (d) ComParE + WEF

Fig. 2: Confusion Matrices on the Evaluation Set

5 Conclusion

In this paper, the performance of temporal, spectral and wavelet features with DNNs was tested and verified for the task of ASC. From this study, we find that WEF has the slightly better performance than ComParE on both of the development and the evaluation set in DCASE 2017 task. WEF achieved 74.8 % and 60.2 % on the development and the evaluation set, respectively. However, the best model in the task is the fusion model of ComParE + WEF with MLP, significantly outweighed the baseline system according to the result of a z-test. The accuracy on the development set was 79.2 % for the fusion model and 74.8 % for the baseline. On the evaluation set, the fusion model achieved 63.8 % which beat the performance of the baseline (61.0 %). This systems performance is competitive, yet in the future, Deep Belief Networks (DBNs), Stacked Autoencoders and GANs could be applied to optimise the result. Besides, utilising low-dimension features with a much deeper network structure would worth investigation.

Acknowledgement

This work was partially supported by the Horizon H2020 Marie Skłodowska-Curie Actions Initial Training Network European Training Network (MSCA-ITN-ETN) project under grant agreement No. 766287 (TAPAS) and the Bavarian State Ministry of Education, Science and the Arts in the framework of the Centre Digitisation.Bavaria (ZD.B).

References

1. F. Eyben, F. Weninger, F. Gross, and B. Schuller. Recent developments in opensmile, the munich open-source multimedia feature extractor. In *Proc. ACM MM*, pages 835–838, Barcelona, Catalunya, Spain, 2013.
2. K. Qian, C. Janott, Z. Zhang, C. Heiser, and B. Schuller. Wavelet features for classification of vote snore sounds. In *Proc. ICASSP*, pages 221–225, Shanghai, China, 2016.
3. K. Qian, C. Janott, V. Pandit, Z. Zhang, C. Heiser, W. Hohenhorst, M. Herzog, W. Hemmert, and B. Schuller. Classification of the excitation location of snore sounds in the upper airway by acoustic multifeature analysis. *IEEE Transactions on Biomedical Engineering*, 64(8):1731–1741, 2017.
4. Alice Baird, Shahin Amiriparian, Nicholas Cummins, Alyssa M Alcorn, Anton Batliner, Sergey Pugachevskiy, Michael Freitag, Maurice Gerczuk, and Björn Schuller. Automatic classification of autistic child vocalisations: A novel database and results. In *Proc. of INTERSPEECH*, pages 849–853, Stockholm, Sweden, 2017.
5. A. Mesaros, T. Heittola, and T. Virtanen. TUT database for acoustic scene classification and sound event detection. In *Proc. EUSIPCO*, pages 1128–1132, Budapest, Hungary, 2016.

6. Marc C Green and Damian Murphy. Acoustic scene classification using spatial features. In *Proc. DCASE Workshop*, pages 42–45, Munich, Germany, 2017.
7. D. Barchiesi, D. Giannoulis, D. Stowell, and M. D. Plumbley. Acoustic scene classification: Classifying environments from the sounds they produce. *IEEE Signal Processing Magazine*, 32(3):16–34, 2015.
8. J. T. Geiger, B. Schuller, and G. Rigoll. Large-scale audio feature extraction and svm for acoustic scene classification. In *Proc. WASPAA Workshop*, pages 1–4, New Paltz, NY, USA, 2013.
9. A. Mesaros, T. Heittola, A. Diment, B. Elizalde, A. Shah, E. Vincent, B. Raj, and T. Virtanen. Dcase 2017 challenge setup: tasks, datasets and baseline system. In *Proc. DCASE Workshop*, pages 85–92, Munich, Germany, 2017.
10. Stephane Mallat. *A wavelet tour of signal processing: the sparse way.* Elsevier, Burlington, MA, USA, 2009.
11. Haibo He and Janusz A Starzyk. A self-organizing learning array system for power quality classification based on wavelet transform. *IEEE Transactions on Power Delivery*, 21(1):286–295, 2006.
12. Andrew Keong Ng, Tong San Koh, Udantha Ranjith Abeyratne, and Kathiravelu Puvanendran. Investigation of obstructive sleep apnea using nonlinear mode interactions in nonstationary snore signals. *Annals of Biomedical Engineering*, 37(9):1796–1806, 2009.
13. David Li, Jason Tam, and Derek Toub. Auditory scene classification using machine learning techniques. 2013.
14. K. Qian, Z. Ren, V. Pandit, Z. Yang, Z. Zhang, and B. Schuller. Wavelets revisited for the classification of acoustic scenes. In *Proc. DCASE Workshop*, pages 108–112, Munich, Germany, 2017.
15. M. Valenti, S. Squartini, A. Diment, G. Parascandolo, and T. Virtanen. A convolutional neural network approach for acoustic scene classification. In *Proc. IJCNN*, pages 1547–1554, Anchorage, AK, USA, 2017.
16. Z. Ren, V. Pandit, K. Qian, Z. Yang, Z. Zhang, and B. Schuller. Deep sequential image features for acoustic scene classification. In *Proc. DCASE Workshop*, pages 113–117, Munich, Germany, 2017.
17. Seongkyu Mun, Sangwook Park, David K Han, and Hanseok Ko. Generative adversarial network based acoustic scene training set augmentation and selection using SVM hyper-plane. In *Proc. DCASE Workshop*, pages 93–102, Munich, Germany, 2017.
18. G. Hinton, L. Deng, D. Yu, G. E. Dahl, A. r. Mohamed, N. Jaitly, A. Senior, V. Vanhoucke, P. Nguyen, T. N. Sainath, and B. Kingsbury. Deep neural networks for acoustic modeling in speech recognition: The shared views of four research groups. *IEEE Signal Processing Magazine*, 29(6):82–97, 2012.
19. D. Ciregan, U. Meier, and J. Schmidhuber. Multi-column deep neural networks for image classification. In *Proc. CVPR*, pages 3642–3649, Providence, RI, USA, 2012.
20. Jee-Weon Jung, Hee-Soo Heo, IL-Ho Yang, Sung-Hyun Yoon, Hye-Jin Shim, and Ha-Jin Yu. Dnn-based audio scene classification for dcase 2017: Dual input features, balancing cost, and stochastic data duplication. In *Proc. DCASE Workshop*, pages 59–63, Munich, Germany, 2017.
21. Florian Eyben. *Real-time speech and music classification by large audio feature space extraction.* Springer, Switzerland, 2015.
22. Ronald R Coifman, Yves Meyer, and Victor Wickerhauser. Wavelet analysis and signal processing. In *Wavelets and their Applications*, pages 153–178, Sudbury, 1992. MA: Jones and Barlett.

23. Rami N. Khushaba. *Application of Biosignal-driven Intelligent Systems for Multi-function Prosthesis Control.* University of Technology Sydney, Sydney, Australia, 2010. Doctoral Thesis.
24. Rami N Khushaba, Sarath Kodagoda, Sara Lal, and Gamini Dissanayake. Driver drowsiness classification using fuzzy wavelet-packet-based feature-extraction algorithm. *IEEE Transactions on Biomedical Engineering*, 58(1):121–131, 2011.
25. M.W Gardner and S.R Dorling. Artificial neural networks (the multilayer perceptron)—a review of applications in the atmospheric sciences. *Atmospheric Environment*, 32(14):2627–2636, 1998.
26. M. L. Seltzer, D. Yu, and Y. Wang. An investigation of deep neural networks for noise robust speech recognition. In *Proc. ICASSP*, pages 7398–7402, Vancouver, BC, Canada, 2013.
27. S. K. Pal and S. Mitra. Multilayer perceptron, fuzzy sets, and classification. *IEEE Transactions on Neural Networks*, 3(5):683–697, 1992.
28. Tobias Scheffer, Christian Decomain, and Stefan Wrobel. Active hidden markov models for information extraction. In *Proc. IDA*, pages 309–318, Cascais, Portugal, 2001.
29. Martín Abadi, Ashish Agarwal, Paul Barham, Eugene Brevdo, Zhifeng Chen, Craig Citro, Greg S. Corrado, Andy Davis, Jeffrey Dean, Matthieu Devin, Sanjay Ghemawat, Ian Goodfellow, Andrew Harp, Geoffrey Irving, Michael Isard, Yangqing Jia, Rafal Jozefowicz, Lukasz Kaiser, Manjunath Kudlur, Josh Levenberg, Dan Mané, Rajat Monga, Sherry Moore, Derek Murray, Chris Olah, Mike Schuster, Jonathon Shlens, Benoit Steiner, Ilya Sutskever, Kunal Talwar, Paul Tucker, Vincent Vanhoucke, Vijay Vasudevan, Fernanda Viégas, Oriol Vinyals, Pete Warden, Martin Wattenberg, Martin Wicke, Yuan Yu, and Xiaoqiang Zheng. TensorFlow: Large-scale machine learning on heterogeneous systems, 2015. Software available from tensorflow.org.
30. Murray R Spiegel, John J Schiller, R Alu Srinivasan, and Mike LeVan. *Probability and Statistics.* McGraw-Hill, New York, NY, USA, 2009.

MIX
Papier aus verantwortungsvollen Quellen
Paper from responsible sources
FSC® C105338

If you have any concerns about our products,
you can contact us on
ProductSafety@springernature.com

In case Publisher is established outside the EU,
the EU authorized representative is:
Springer Nature Customer Service Center GmbH
Europaplatz 3, 69115 Heidelberg, Germany

Printed by Libri Plureos GmbH
in Hamburg, Germany